MODERN RIVERSIDE SCENE AT QINGMING FESTIVAL
Positioning, Planning, Design & Operation of Cultural and Commercial Street

现代清明上河图
——文化商业古街的定位、规划、设计、运营全解析

唐艺设计资讯集团有限公司 策划
广州市唐艺文化传播有限公司 编著

图书在版编目（CIP）数据

现代清明上河图：文化商业古街的定位、规划、设计、运营全解析：全3册 / 广州市唐艺文化传播有限公司编著. — 天津：天津大学出版社，2014.1
ISBN 978-7-5618-4933-0

I.①现…II.①广…III.①商业街—城市规划—中国②商业街-建筑设计-中国 IV.①TU984.13

中国版本图书CIP数据核字（2014）第012847号

责任编辑　朱玉红
装帧设计　林国代
文字整理　陈　建　樊　琼　许秋怡　王艳丽
流程指导　王艳丽
策划指导　黄　静

现代清明上河图——文化商业古街的定位、规划、设计、运营全解析

出版发行　天津大学出版社
出 版 人　杨欢
地　　址　天津市卫津路92号天津大学内（邮编：300072）
电　　话　发行部022-27403647
网　　址　publish.tju.edu.cn
印　　刷　利丰雅高印刷（深圳）有限公司
经　　销　全国各地新华书店
开　　本　245mm×325mm　1/16
印　　张　50.5
字　　数　926千
版　　次　2014年4月第1版
印　　次　2014年4月第1次
定　　价　838.00元（USD 180.00）

Preface
序

Regional districts owning certain cultural deposits or traditional features are injected with cultural, leisurely and creative elements, so as to be renovated, extended or reconstructed as "international, cultural and fashionable" leisure commercial streets.

In recent years, after commercial real estates' rapid expansion, cultural commercial streets, with specific historical and cultural background, specific consumers and rich business activities, and combined with traditional culture, history and modern civilization, are popular in commercial real estate market. Their new development mode receives great popularity and is identified as a more intimate and sustainable mode.

With great popularity, domestic cultural commercial streets emerge in large numbers. On one hand, domestic commercial real estate magnates such as Wanda Group and Shui On Group all spare no expense to develop cultural commercial districts. On the other hand, unique historical and cultural deposits differentiate the cultural commercial districts from traditional commercial streets and form their own developing features and operation modes. In such a situation we edit this book themed with commercial street market positioning, planning and design, and operation, which includes many successful cultural commercial district projects on market, so as to provide all developers and designers interested in cultural commercial streets development with good examples for reference.

This publication includes three volumes with 44 representative cases collected from 30 cities of 18 provinces and municipalities in China. All cases are of high quality and wide scope, with high-definition pictures provided by our professional photographers on the spot.

According to development characteristics, the book is divided into 4 categories, i.e. commercial streets reconstructed from ancient residential areas and ancient streets, commercial streets extended from ancient towns, modern imitated commercial streets of ancient style, and commercial streets reconstructed on the basis of historic preservations, to present the development status and planning design features of domestic cultural commercial streets from all aspects.

There are 7 sections to introduce each case. From street background and market positioning, planning, design features, commercial activities and operation, to brand shops and cultural facilities, all detail descriptions are arranged for holistic presentation of the development and operation process of each case to provide general developers and designers with new and practical references.

　　文化商业街区，是指在有一定文化底蕴或传统特色的区域地段，注入"文化、休闲、创意"元素，改造、扩建或仿建出具有"国际性、文化性、时尚性"的休闲娱乐商业街区。

　　近年来，商业地产在经历同质化放量增长阶段后，文化商业街区这种以特定的历史文化为背景、以特定的消费人群为导向、以丰富的时尚业态为支撑、融合传统历史文化与现代文明的商业街区，受到商业地产市场的青睐。这种新型的商业地产开发模式受到业界的追捧，被业界认定为一种更具亲和力和可持续发展的经典模式。

　　在商业地产市场和业界的双重青睐下，国内文化商业街区开发日趋火热。一方面，国内商业地产大鳄如万达集团、瑞安集团、1912集团等纷纷斥巨资打造文化商业街区项目，商业地产市场上涌现出了一批成功的文化商业街区开发案例，如佛山岭南新天地、宽窄巷子、楚河汉街、上海新天地、南京1912等；另一方面，由于文化商业街区独特的历史文化底蕴，使文化商业街区的开发理念与开发模式有别于传统的商业街区，形成自己的开发特点、运营模式。在此种局势下，我们策划了此套以文化商业街区的定位、规划、设计、运营为主题的图书，收录市场上成功的文化商业街区开发项目，旨在为有志于文化商业街区开发的开发商与设计师提供参考和借鉴。

　　《现代清明上河图》分上、中、下三册，共收录北京、天津、上海、重庆、山东、江苏、浙江、福建、广东、湖北、湖南、四川等18个省市30个城市的44个代表性项目，项目质量高、辐射范围广、数量大，项目图片全部由专业摄影师实地拍摄。

　　本书中，我们将项目按开发特点划分为四大类，分别是古住宅和古街道改建的商业街区、古城镇扩建的商业街区、现代仿建的商业街区和以文物保护为重点改建的商业街区，全方位展现我国文化商业街区的开发现状和规划设计特色。

　　在每个街区的编排中，我们从街区背景与定位、街区规划、设计特色、商业业态、市场运营、品牌商铺、文化设施七个方面详细介绍，力求全面呈现项目开发运营全过程，为广大开发商和设计师提供鲜活而极具实用性的文化商业街区参考案例。

CONTENTS 目录

022-262

现代仿建的商业街区
Modern Imitated Commercial Streets of Ancient Style

024	Impression of 1912	印象1912
026	1912 Nanjing	南京1912
054	1912 Hefei	合肥1912
064	Chu River & Han Street, Wuhan	武汉楚河汉街
078	Chenghuang Temple Commercial Street, Shanghai	上海城隍庙商业街
098	Yancheng Chunqiu Tourist Area, Changzhou	常州淹城传统商业街坊
112	Ligongdi, Suzhou	苏州李公堤
130	Ancient Culture Street, Tianjin	天津古文化街
154	Imperial Song Street, Kaifeng	开封宋都御街
166	Xibu Alley, Zhangjiajie	张家界溪布街
184	Lingnan Impression, Guangzhou	广州岭南印象园
208	Litchi Bay, Guangzhou	广州荔枝湾
234	Wenfeng Gujie, Chongqing	重庆文峰古街
246	Commercial Pedestrian Street, Beichuan	北川商业步行街
256	China Qiangcheng, Wenchuan	汶川中国羌城

现代清明上河图
——项目经典片段摘录

所有的名堂，全部青墙红柱、磨砖对缝，修饰以形式各样的隔扇门窗、栏杆、屋顶翼角，古朴、典雅又不失隽秀。匾额、楹联、宫灯、旗幡、精细的木雕及上千幅色彩艳丽的油漆绘画，使街道弥漫着古典的文化气息。街路以『十二铜钱』『十二生肖印章』铺装，配合灯光效果设计，使得商贸区呈现古韵新风。

——天津古文化街

项目采用『化整为零』的设计策略，在旧有的城市肌理完整的地区分散大体量建筑，避免传统的城市肌理与大体量建筑产生矛盾。通过整体性控制所有的建设过程，使一座单体建筑与其他单体建筑彼此兼顾，链锁式发展，为恢复城市空间的传统意蕴和延续城市记忆的城市空间提供保证。

——上海城隍庙商业街

常州淹城传统商业街坊把春秋的主体文化、最重要的文化、流传至今的典故

古住宅和古街道改建的商业街区

Commercial Streets Reconstructed from Ancient Residential Areas and Ancient Streets

022-339

- 026 佛山岭南天地 Lingnan Tiandi, Foshan
- 052 武汉天地 Wuhan Tiandi
- 182 黄山屯溪老街 Tunxi Street, Huangshan
- 248 苏州观前街 Guanqian Street, Suzhou

上

- 074　上海新天地　Xintiandi, Shanghai
- 100　北京前门大街　Qianmen Street, Beijing
- 200　黄山置地国际中心　Huangshan Land International Center
- 258　扬州东关街　Dongguan Street, Yangzhou
- 272　宁波南塘老街　Nantang Street, Ningbo
- 288　福州三坊七巷　Sanfang Qixiang, Fuzhou

转化为观光、游乐、互动等体验项目，如用孔子学堂的方式来展现儒家文化、用雕刻绘画的形式在山体中展现富有传奇色彩的春秋丽人的故事场景、用春秋版图的形式表现春秋历史地理的文化内涵、用歌舞升平来包装旋转木马、用声光水影效果演绎主题水影秀等。

——常州淹城传统商业街坊

通过『桥堤文化』和『湖滨公园』把金鸡湖的水、绿与姑苏的文化结合在一起，将金鸡湖与现代多元风情、历史与现实、休闲旅游与商业有机地组合起来。设计以挖掘与引入并重，不仅对李公堤历史人文进行了深入挖掘，完成了李公堤碑亭、李超琼书画、李超琼雕塑及诗碑、李公堤记碑文等文化景观的营造，还引入多元个性文化，为街区文化不断『输入灵魂』。

——苏州李公堤

在建筑外观上，毫无修饰的青砖既是墙体，又是外部装饰。烟灰色的墙面上，勾勒了白色的砖缝，除此之外再无任何修饰。这片青灰色与砖红色相间的建筑群，风格古朴精巧，错落有致地呈『L』形环绕『总统府』，成为以民国文化为建筑特点的商业建筑群。

——南京1912

设计引入徽派文化，将现代和传统进行融合。在空间处理上，以一条主题商业内街为主轴，将沿黄山路的高档餐饮区和内部的情景合院式酒吧区串联，电影院、精品酒店、娱乐中心、经济型酒店穿插其中，既形成较为醒目的地标，亦成为情景商业的绚丽背景。历史与明天、文化与商业在此邂逅，和谐而统一。

——合肥1912

设计以东京梦华录为蓝本，大小店铺均采用宋代营造方式，全街南高北低，多为二三层建筑，既有三步两店的一般店铺，又有体量宏大的仿古楼阁。沿街建筑青砖灰瓦，褐柱画梁，朱栏雕窗，主要建筑还施以宋式彩绘，使古都开封成为旅游开放城市中颇具魅力的一站。

——开封宋都御街

汉街主体采用民国建筑风格，红灰相间的清水砖墙、精致的砖砌线脚、乌漆大门、铜制门环、石库门头、青砖小道、老旧的木漆窗户，置身其中，仿佛时光倒流。同时，汉街中又将时尚的现代建筑和欧式建筑穿插在民国风格建筑中，实现传统与现代的完美融合。

——武汉楚河汉街

采用丰富的空间语言进行建筑的平面功能设计，从街道、入户至庭院都尽可能地留有一定的空间冗余量，以丰富建筑空间的形态和体验感。在立面造型上，设计师对土家族和苗族的传统语言进行艺术再加工和简洁的抽象变形，将其作为建筑表现的构成元素，使立面风格统一、古朴大气。

——张家界溪布街

园内民居依水而建，或窄门高屋，或镬耳高墙。悠长的青云巷、古朴的趟栊

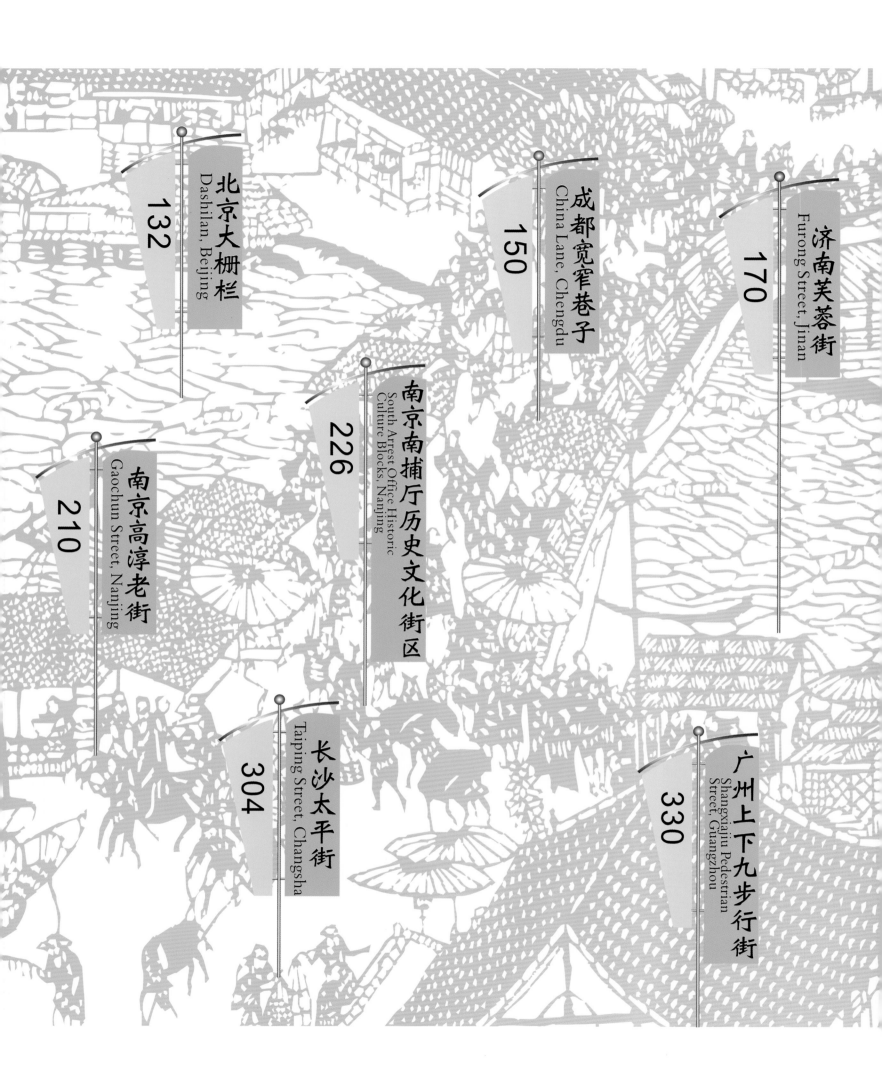

现代仿建的商业街区
Modern Imitated Commercial Streets of Ancient Style

022-262 中

- 026 南京1912 / 1912 Nanjing
- 054 合肥1912 / 1912 Hefei
- 064 武汉楚河汉街 / Chu River & Han Street, Wuhan
- 112 苏州李公堤 / Ligongdi, Suzhou
- 130 天津古文化街 / Ancient Culture Street, Tianjin
- 208 广州荔枝湾 / Litchi Bay, Guangzhou
- 234 重庆文峰古街 / Wenfeng Gujie, Chongqing

门、精致的满洲窗，小溪蜿蜒，池塘清澈，散发出岭南水乡的韵味。园内有包拯相祠、敬佛堂、天后宫等民间祭拜文化区，又有老酒坊、老理发店、老电影院、老照相馆等令人怀旧的场所，还有新会葵艺馆、广绣馆、木雕宫灯等民族工艺展示区，让游客时刻感受到浓浓的岭南风情。

——广州岭南印象园

石脚水磨青砖的西关大屋、优雅的花园式别墅、名人公寓、中西合璧的骑楼等，这些不同时期、不同风格的街区和建筑在一定程度上反映了广州各个发展时期的历史和文化轨迹，形成了古朴典雅、中西交融、岭南文化气息浓郁的城市风貌。

——广州荔枝湾

在建筑布局上，充分运用中国传统风水学，整体布局错落有致、进退有序，以回廊、下行通道、临江步行栈道构建起了立体多维的交通体系和人流体系，使整个商业街区可以随意贯穿、自由通达，既符合古代街市内聚人气的传统特色，又承袭现代开放式步行街集聚人流的时尚商业形态。

——重庆文峰古街

古风遗韵与现代元素相融合的设计，使历经千百年的羌族民居建筑焕发出新的光彩。建筑的每一处檐角、女儿墙顶、门窗都点缀着传统的羌族民俗文化符号——白色羊皮褂、白头帕、白布裙……这些建筑彼此之间通过走廊、天桥和楼梯连接，既独立又连通，自成单元又家家相通。

——北川商业步行街

multi-dimensional circulation system accessible to any places within the program.

— Wenfeng Gujie, Chongqing

The combination of antique charm and modern elements makes the old residential buildings of Qiang irradiate. Every detail of the buildings is dotted with traditional Qiang folk cultural symbol. All the buildings connect with each other by corridors, overpasses, and stairs; therefore the buildings are independent yet connected with each other.

— Commercial Pedestrian Street, Beichuan

The design concept is based on strong local and ethnic feature of Weizhou. Xiqiang Culture Street and Guozhuang Square of Qiang style are built along the Minjiang River. Sepia interspersed with sandy beige is applied, sheep totems are embedded on walls. Columns with sheep totems are standing on the street, and also Qiang traditional watchtowers are adopted.

— China Qiangcheng, Wenchuan

从威州的浓郁地域、民族特色出发，着力建设现代羌城，将『西羌圣城』的理念贯穿于设计的全过程。在岷江一线建设具有羌族建筑风格的西羌文化街和锅庄广场。建筑色彩采用羌族喜爱的片石浅褐色点缀原木棕褐色，建筑墙上嵌入羊图腾，街道上立着有羊图腾的柱子，另外该项目还运用碉楼这一传统羌城建筑元素。

——汶川中国羌城

Confucian culture is presented by Confucian School, story scene of Chunqiu beauties is presented by sculptures and paintings on massifs, and acousto-optic water effects are used to present themed watermark show.

— Yancheng Chunqiu Tourist Area, Changzhou

Ligongdi organically combines the Jinji Lake with modern multi-style, history with reality, leisure tourism with business. The design pays attention to both excavation and introduction, which means to explore the local history and culture, create cultural landscapes, and enrich the culture of the street by importing other characteristic cultures.

— Ligongdi, Suzhou

As for architectural appearance, unadorned black bricks constitute the wall and serve as exterior decorations. The smoky grey wall is only decorated with white brickwork joints. Grey and brick-red buildings alternately arranged show archaic and delicate style and encircle "the Presidential Palace" in an L shape, creating a commercial building complex of cultural flavor.

— 1912 Nanjing

Hui-style culture is brought in to integrate modernity with tradition. A themed commercial street as the axis links up restaurant area and bar area along Huangshan Road with cinema, boutique hotels, recreational center and budget hotels interspersed. Several themed courtyards and squares are formed between streets and buildings to create an outdoor commercial place full of atmosphere. History and future, culture and business meet here harmoniously.

— 1912 Hefei

Based on *Dongjing Menghua Lu*, the design of Imperial Song Street highlights features of Song Dynasty. All shops of different sizes are built in Song style. Two- or three-floor buildings higher at south and lower at north distribute along the street. All the buildings are built of black bricks and grey tiles and are decorated with brown columns, painted beams and carved windows. Major buildings are decorated with colored paintings in Song style.

— Imperial Song Street, Kaifeng

The main body of Han Street mainly adopts building style of Republic of China. When walking on the street, visitors will feel like going back to the old times. Meanwhile, modern buildings and European-style buildings are alternated among the buildings of Republic of China style to achieve a perfect mixture of the tradition and the modernity.

— Chu River & Han Street, Wuhan

Redundancies are retained as far as possible from streets to courtyards to enrich the shape of building space and experience. Traditional languages of Tujia and Miao minorities are applied in façade design after art reprocessing.

— Xibu Alley, Zhangjiajie

The residences are built along the river. The long Qingyun Lane, antique security doors, delicate Manchu windows, wandering stream, and clear pond diffuse a kind of charm of Lingnan water village. There are folk worship cultural areas including Bao Zheng Ancestoral Hall, Tianhou Temple, and also nostalgic places such as old wine shop, old barbershop, old cinema, old photo studio and so on, and traditional craft exhibition zones such as Xinhui Palm-leaf Fan House, Guangzhou Embroidery House and so on.

— Lingnan Impression, Guangzhou

Streets and buildings built in different periods of different styles reflect the history and culture of Guangzhou in different developing stages to some extent, and create a simple and elegant city style of profound Lingnan culture.

— Litchi Bay, Guangzhou

For architectural layout, traditional geomantic theory is fully applied to arrange all buildings with an orderly hierarchy. Corridors, down channels, riverside walkways constitute a

古城镇扩建的商业街区
Commercial Streets Extended from Ancient Towns

022-129

下

024	山西平遥南大街 Pingyao South Street, Shanxi
036	山西朔州老街 Shuozhou Old Street, Shanxi
056	苏州同里 Tongli, Suzhou
108	丽江大研古城 Dayan Old Town, Lijiang

以文物保护为重点改建的商业街区
Commercial Streets Reconstructed on the Basis of Historic Preservations

130-206

- 132 南京夫子庙 Confucius Temple, Nanjing
- 154 镇江西津渡古街 Xijin Ferry Street, Zhenjiang
- 174 绍兴仓桥直街 Cangqiao Zhijie, Shaoxing
- 186 厦门鼓浪屿 Gulangyu Island, Xiamen

Highlights from Selected Projects
Modern Riverside Scene at Qingming Festival

All the buildings are decorated with various partition boards, doors, windows, rails, tablets, lanterns, wood sculptures and colorful paintings. The street is paved with sculptures, and matched with light effect, to make the commercial area take on an ancient taste.

— Ancient Culture Street, Tianjin

Design strategy of "breaking up the whole into parts" was supplied to avoid the contradiction between traditional city texture and large buildings. By integrally controlling all the construction process, each individual building is connected to others, and therefore making sure that the traditional connotation of city space is recovered and memories of the city is inherited.

— Chenghuang Temple Commercial Street, Shanghai

Major culture in the Spring and Autumn Period, culture essence and allusions spreaded so far are transformed into interaction experience project of sightseeing and entertainment. For example,

Modern imitated commercial street is different from cultural relic or historic site projects. It not only records the commercial activity of human beings, but also has basic functions of modern urban life. Therefore, a suitable development mode should be built through interaction of multi-layer layouts, reorganization of multi-perspective relations and integration of operation of multiple dimensions.

Interaction of Multi-layer Layouts

1. Layout of material space. First, to adequately cherish features of imitated buildings and clarify reasonable layouts of projects. Second, to form a circulating space feature due to traffic organization of cities. Third, to pay attention to historical memory.
2. Layout of cultural spirit. To construct a new spirit layout on the view of "macro-history". To deeply explore its cultural wisdom, refine its unique culture characteristic and recall people's collective memory of culture by products designed for appreciating and experiencing. Therefore, to target at a broader market.
3. Production of living function. Tourist function, leisure function and living function are orderly combined to realize double benefits.

Reorganization of Multi-perspective Relations

1. Position relation. The reorganization of new relations among positioning of commercial activities, culture and brand sets the most reasonable development approach for streets.
2. Space relation. Cross-over design of spaces of landscape, function, rest, sightseeing and consumption sets the most comfortable space order for streets.
3. Production relation. Integrated design of "politic production", "dwelling production" and "operation production" sets the most feasible cooperation structure for streets.

Integration of Operation of Multiple Dimensions

1. Cultural operation. Business cultural information of different dynasties is added up to avoid freeze-frame of history and stiff sense of static space, and thus, to form a interesting scenery.
2. Marketing operation. Marketing modes make full use of TV serials, cultural trends and city brands.

With the rapid development of economy, there are more and more projects of imitated commercial street. However, not all these projects can be acknowledged by the public and market. Today, cultural theme and style are the critical factors that decide whether a commercial street can remain invincible in future competition.

现代仿建商业街不同于文物古迹类、古城镇类项目，它不仅记录着人类的商业活动，还具有现代城市生活的基本功能，既是物化的文化遗产，又必须考虑城市建设的现实导向。因此，在开发建设过程中，既不能单纯地从文化角度进行封闭的历史再现式演绎，又不能简单地试图通过商业开发来解决城市发展的问题，而是通过多层次格局的互动、多视角关系的重组、多维度运作的整合，构建出适合自身发展的模式。

多层次格局的互动

1. 物质的空间格局。首先，对被仿照建筑所具有的特点充分地珍视，明晰项目的合理格局；其次，因城市的交通组织、地块关联等因素，街区的内外之间形成循环往复的空间特质；另外，在对空间格局进行研究和规划的同时，注重其所呈现的历史记忆。
2. 文化的精神格局。寻找文化脉络的"基因链"，以"大历史"的视野构建新的精神格局。通过"转基因"建设，深刻挖掘其所折射的文化智慧所在，凝练出特有的文化性格，并通过可观赏、可品味、可体验的产品设计，面向更广泛的市场传达其特有的文化意义及趣味。
3. 生活的功能格局。面向游客的旅游功能、面向区域公众的休闲功能、面向本地居民的生活功能等有秩序地结合在一起，通过良性的功能互动来实现效益的叠加。

多视角关系的重组

1. 定位关系。对业态定位、文化定位、品牌定位等进行新的关系组织设计，为街区设定最合理的发展途径。
2. 空间关系。对景观空间、功能空间、游憩空间、游览空间、消费空间等进行交叉设计，为街区设定最舒适的空间秩序。
3. 生产关系。对政府的"政策型生产"、当地居民的"生活型生产"、旅游经营者的"经营型生产"等进行结合设计，为街区设定最可行的合作结构。

多维度运作的整合

1. 文化运作。通过时间符号的传递式叠加，如在唐、宋、元、明、清等历史符号的基础上，叠加新中国成立初期的商业文化信息，规避历史定格和静态空间的生硬感，形成颇具玩味的生动景象。
2. 营销运作。营销运作方式包括借视营销（借助电视剧制作和热播）、借势营销（借助市场对商业智慧及文化精神传承越来越重视的趋势）和借市营销（依托城市品牌发展导向进行大规模宣传）。

随着经济的快速发展，国内出现了越来越多的仿建商业街项目，但并不是所有的项目都能得到民众和市场的认可。"物相杂，则文生；文不正，则乱成。" 这说明，在仿古泛滥的今天，文化主题与文化风格决定了一个商业街是否具有品牌优势，是否能在将来的竞争中立于不败之地。

Impression of 1912
印象1912

"1912" is a business form, and its core commercial activities cover bar, restaurant, recreation, entertainment, cinema and hotel. However, unlike common commercial programs, it focuses on using creative ideas to achieve an integrated operation of "city, culture and commerce". It creates spirit of the place in coincidence with city cultural temperament through architecture and, through exploration and search of coexistence, it finishes the evolution from architectural culture to city taste and consumption culture. It is generally embodied at the following three aspects:

City Parlor

Due to its consumer-driven property, full-time cluster form, profound city history and cultural deposits, 1912 becomes a unique parlor of each city, and the premier place for city white-collar workers and the leisure class to have relaxation and banquets. It is also an ideal place for business negotiation and communication, and is the hot spot region for launching city fashion events.

Cultural Card

1912 makes deep research and judgment based on local cultural deposits, business development degree and consumption habits. Through different business proportions and unique operation mode, each carefully-orchestrated 1912 becomes the "loudspeaker" of the city culture, which not only expands the channels of city culture dissemination, but also improves the city popularity.

Fashion Landmark

1912 which insists on high-end culture experience consumption pattern pays much attention to the entire program image as well as its quality. Therefore, it establishes a strategic alliance with domestic and international famous brands, so as to create a quality brand for the city. In every 1912 branch, "Celebrities, boutiques, top stores and well-known exhibitions" are used as the bond to strengthen its position as the city No.1 fashion landmark, according to consumers' characteristics.

Strategic Location
Targeted Cities

As a medium and high-end fashional and leisure consumption block, 1912, whose location pays much attention to the city's economic level and consumption culture, prefers to choose the city with certain economic base and great development potential. There are mainly three kinds of cities included: nationwide municipalities, provincial capitals and listed planning cities with developed economy; prefectures within Jiangsu Province and coastal second-tier cities of high economic level outside Jiangsu Province; well-developed counties at Yangtze River Delta and Pearl River Delta.

Targeted Regions

Apart from the attention paid to the targeted cities, 1912, as a commercial block leading city fashion consumption trend, is also strict with its region choice.

Geographical Position — For projects in expanded areas, the choice of site should consider the business development level and consumption potential of the region. For the main downtown program, areas within 3 km around the core business circle are preferred locations, followed by the sub-center or the border of sub-center and core business circle. For the programs in new urban areas, regional business centers are the major consideration of the site.

Traffic — Traffic should be convenient at the site. Smooth city roads and circulation within the program will help attract consumer groups.

Cultural Features — The organic combination of city culture and business forms should be valued. Whether the place owns cultural features and historical relics is one of the important consideration standards. Regions with certain historical background and cultural relics are key consideration of those expanded areas.

Ways of Cooperation
Programs with commercial properties

In principle, independent land ownership and only operation mode will ensure the ownership of the block. Such practices will not only ensure that the program operate independently and professionally, but also keep each block working continually, which helps create a business and cultural card to the city.

Programs with land or with expected land

The form of joint venture should be used to acquire land. At the same time, a management company should be founded by 1912 Group and be entrusted with the assets to manage and operate the entire block. The company should participate in the plan, design, investment and other jobs from the initial phase to ensure the smooth operation of the project after completion and realize the interests of all.

Programs with commercial or residential properties

High-quality programs will be introduced into core areas of the first- and second-tier cities and operated by 1912. The period of entrusting assets to 1912 should be no less than 15 years.

Status of Operation
1912 Nanjing

Located at the intersection of Changjiang Road and Taiping North Road, 1912 Nanjing is comprised of 19 buildings of architectural style of the Republic of China, as well as four squares named Republic, Charity, New Century and the Pacific. Centered on the "Nanjing Presidential Palace" in L shape, covering 42,000 m² of land, the project can be divided into three sections: leisure area, bar area and restaurant area. So far the program has attracted more than 50 famous brands at home and abroad. It is one of the best places for leisure consumption and there are over 30,000 passengers per day.

1912 Suzhou

Located on the Ligongdi within Suzhou Industrial Park, 1912 Suzhou borders Jinji Lake on the north and Suzhou River on the south, and covers 13,000 m² of land. With charming European-style buildings smartly combined with neoclassical style pavilions and bridges, the project becomes the first waterfront international commercial street integrating dining, bar, recreation and entertainment. It is the most dazzling "luminous pearl" and the new entertainment landmark in Suzhou.

1912 Wuxi

1912 Wuxi located in the shopping mall in front of the railway station covers about 13,000 m² of land. Guided by a theme of "bar lounge", it optimizes the dispersive and single distribution of the bars in Wuxi and reflects the management ability of the brand program over specific commercial activities. It opens a new era of the city's night life, fills the past lack of business forms in Wuxi and becomes the card of the city fashion taste.

1912 Yangzhou

1912 Yangzhou is located at southeast in Yangzhou, next to the ancient canal. It is composed of Lu Family's ancient houses and seven buildings of Ming and Qing dynasties style. Covering 34,000 m², the project can be divided into three parts including bar, featured dining and recreation. Many famous brands at home and abroad are attracted to Yangzhou, together with local brands, to create a high-quality leisure consumption land.

"1912"是一种商业形态，核心业态主要有酒吧、餐饮、休闲、娱乐、影院、酒店。但是，它又不局限于一般性商业项目的思考，其核心在于以创意的方法，达到"城市——文化——商业"的整合运营。通过建筑所营造的场所精神与城市文化气质相吻合，在探索与寻求和谐的过程中，完成建筑文化到城市品位与消费文化的涅槃，具体体现在以下三个方面。

城市客厅

"1912"因其消费的导向性、全天候集群业态、城市历史文化的沉淀，成为各城市独一无二的城市客厅，成为城市白领和小资阶层休闲宴请的首选之地、本地及外地的中外商务人士洽谈与交际的首选场所、城市时尚活动发布的热点区域。

文化名片

"1912"根据当地的文化积淀、商业发展程度及消费习惯进行深度研究和判断，通过不同的业态配比、个性化街区等运营模式，精心打造专属于每个城市的"1912"，成为城市文化的"扬声器"，拓宽了城市文化传播渠道，提升了城市知名度。

品位地标

"1912"坚持高端文化体验式消费路线，注重街区整体形象和品质，通过与大批国际、国内著名商家组成战略联盟，实现强强联合，为城市塑造品牌高地。所到之处，以"名人、名品、名店、名展"为纽带，根据消费客群特征，不断巩固城市时尚"第一名片"的地位，持续造就时尚"注意力"，牢固掌握时尚"话语权"。

战略选址

目标城市

"1912"作为中高端时尚休闲消费街区，其选址十分注重所在城市的经济水平和消费文化，侧重选择有一定经济基础和较大发展潜力的城市，主要包括三类城市：第一类城市——经济发达的直辖市、省会城市、计划单列市；第二类城市——江苏省内地级市以及江苏省外部分经济水平较高的沿海、沿江、拥有地铁的二线城市；第三类城市——长三角、珠三角经济较为发达的县级市。

目标区域

除了对城市的选择，作为引领城市时尚消费潮流的商业街区，"1912"对所在城市的区域也有自身的要求。

区域——异地拓展区域的选择，主要考虑区域内的商业发展水平和消费潜力。对于主城区内的项目，首选入驻城市核心商圈位置3千米范围以内；其次为入驻区域副中心或是副中心与主核心商圈交界处。对于城市新城区内的项目，主要考虑入驻新城区的商业中心区域。

交通——所在地块项目周边车辆通行便捷，外部城市道路和项目内车流动线的组织流畅，便于街区"落客"。

文化特色——重视当地的城市文化与商业业态的有机融合，是否拥有文化特色与历史遗风是选址的重要考量标准之一。拥有一定历史背景、文化遗存的区域是重点考虑的拓展区域。

合作方式

含商业地块的项目

原则上在一二线城市新城核心区域及三线城市采取独立拿地、独立运作的模式拓展新项目，长期持有街区物业，这种方式既保证了项目运作的专业性和独立性，也保证每个街区都能持续性经营，实现打造城市商业和文化名片的目的。

已经取得或即将取得土地的项目

此类项目采用合资的方式共同拿地运作，同时由"1912"成立管理公司，以资产托管方式对街区进行整体运营，在项目起始阶段即参与规划、设计、招商等工作，保证项目建成后的顺利运营，以实现各方利益。

已建商业、物业的项目

在一二线城市核心区域以资产托管的方式进驻较为优质的已有项目，合作方将资产托管于"1912"不低于15年，由"1912"进行运作。

运营现状

南京1912

南京1912位于南京长江路与太平北路交会处，由19栋民国风格建筑及"共和""博爱""新世纪""太平洋"四个街心广场组成，以L形环绕在"总统府"周围，总建筑面积42 000平方米。项目总体分为休闲区、酒吧区、餐饮区三大块，目前已吸引50余家国际、国内品牌商家进驻，成为南京外籍人士、商务人群、高端白领、城市小资的首选休闲消费地，日客流量超过3万人。

苏州1912

苏州1912位于苏州工业园区李公堤，北临中国最大的内城湖泊金鸡湖，南接苏州景观内河，总建筑面积13 000平方米。项目以独具魅力的欧陆风尚建筑与新古典主义道桥亭台巧妙结合，成为苏州首条集餐饮、酒吧、休闲、娱乐为一体的国际风情商业水街，成为苏州最为璀璨的"夜明珠"和娱乐新地标。

无锡1912

无锡1912位于无锡火车站站前商贸城，总建筑面积约13 000平方米。项目以"酒吧休闲长廊"为主题，改善了无锡酒吧分散、单调的格局，体现了品牌项目的业态"统治力"，开启了无锡夜生活的新纪元，成为填补无锡城市商业形态空白、代言无锡城市时尚品位的"名片"。

扬州1912

扬州1912位于扬州城东南，临古运河大水湾畔，由卢氏盐商古宅及七栋仿明清建筑组成，总建筑面积34 000平方米。项目总体分为酒吧娱乐、特色餐饮、休闲娱乐三个区域，众多国际、国内知名品牌借助项目登陆扬州，与扬州当地品牌商家共同打造休闲消费高地。

1912 Nanjing
南京1912

Street Background & Market Positioning
街区背景与定位

History 历史承袭

Nanjing in the time of the Republic of China has gathered the most prominent politicians and scholars and is a place where the East meets the West. Affected by the western style, buildings and social customs of that time also showcase a merged style.

1912 Nanjing is named after the gorgeous fragments of the city's history. Its alluring charm of that time leads modern people into the history of the city.

民国时期的南京城聚集着显赫的政界要人和学术大家，是中西交会之地。受西风东渐的影响，民国时期的建筑、社会风尚都带有中西合璧的味道。

"南京1912"的项目以南京近代史上最辉煌的历史片段命名，让人们领略最令人回味的"民国"情调，带领现代人走进一座城市的历史。

Location 区位特征

1912 Nanjing located at the intersection of Changjiang Road and Taiping North Road borders the Zhujiang Road on the north and is only 1,000 m away from Xinjiekou, China's first business circle. Located at the Daxinggong District, it is adjacent to the famous historical attraction–Nanjing Presidential Palace and overlooks the Nanjing Library across a road. It enjoys prominent business advantages as well as favorable natural and cultural environment.

"南京1912"位于南京市长江路与太平北路交会处，南临南京文化一条街——长江路，北靠南京电子一条街——珠江路，离华夏第一商圈——新街口仅1 000米，处于大行宫地区，紧临历史名胜风景区"总统府"，与南京市图书馆仅一路之隔，尽享核心商圈先天区位优势及周边得天独厚的自然、人文环境。

Market Positioning 市场定位

The project is positioned as a "City Parlor" which shows the cultural essence of the city and gorgeous history of the Republic of China and leads the trend of fashion. Historical flavor and fashion are combined here to create a leisure consumption place merging the East and the West, the fashion and the culture, and a comprehensive leisure block gathering dining, recreation and entertainment.

The project's theme is recreation, fashion and tourism. Its prominent location with large land of few buildings and broad vision contributes to a tranquil open space of interactive atmosphere in the busy streets. It is an unrivalled spot compared with other fashional consumption places.

"南京1912"定位为浓缩南京城市人文精华和民国历史风采并能引领时尚的"城市客厅"，将浓郁的民国历史风情和时尚结合，打造出一个中西合璧、时尚互融、文化精彩的休闲消费场所，集餐饮、休闲、娱乐为一体的综合性时尚休闲街区。

项目主题是休闲、时尚，同时兼具旅游的功能，其最大的优势在于地处黄金地段，面积大，建筑少，视野开阔，形成了一个闹中取静的开放空间以及"里面的人想到外面来，外面的人想到里面去"的互动氛围，这是其他时尚消费场所不具备的独一无二的资源。

规划设计特色
Planning & Design Features

Street Planning 街区规划

A courtyard framework connects all buildings in the street structure and squares are set at key points where new buildings complement the old ones. The project movement line shows as "井"-shaped or "田"-shaped double circulation. The main circulation encircles the entire building complex externally, while the secondary circulation runs around the project internally.

The entire project comprises four squares and three streets, all these feature open, experiential and complex layout. The three streets refer to the external street along the Taiping North Road and the Changjiang Back Street, the internal street along the bounding wall of the Presidential Palace and alleys, the street and central pedestrian blocks between the two. The weaving of the streets and building volumes is conducive to the development and operation of commercial activities.

The program occupying nearly 40,000 m² land covers bar, dining venues and recreation spaces. There are four squares and each block is equipped with over 140 underground parking lots and nearly 100 others on the ground. The first phase project is launched on December 24, 2004. The commercial planning of the second phase will complement and perfect the former, with dining as the key and businesses strictly selected for middle- or top-grade marketing positioning.

街区规划以合院为基本骨架，用街的方式将各幢建筑串联起来，并在关键节点处整合出一些广场空间，新旧建筑相辅相融。其动线成井字形或者田字形双动线，主动线围绕整个建筑群外部，次动线围绕整个建筑群内部，简单明了。

街区主要由4个广场和3条街巷构成。3条街巷分别是沿太平北路及长江后街的外街、沿总统府围墙设置的内街、位于两者中间的中心步行街区和与之相交的一些巷道或街区。广场和街巷的布置为开放性、体验性、复杂性的格局，同时多条街巷与建筑实体的交织也更有利于商业的开发与运营。

建筑总面积约4万平方米，分为酒吧区、餐饮区和休闲区，有共和、博爱、新世纪、太平洋4个广场，街区配备140余个地下停车位及近百个地面停车位。项目一期于2004年12月24日开街，二期的商业业态规划主要对一期进行补充和完善，主打是餐饮，招商坚持中、高档定位，对商家进行严格筛选。

Street Design Features 街区设计特色

The Presidential Palace gathers the most buildings of the Republic of China and so the 1912 Nanjing which relies on it shows the architectural spirit of that time. There are 5 original buildings in the 17 houses, the highest are only three-storey houses and most of them are two-storey or even one-floor houses. To reserve the original look and offer the citizens convenient leisure landscape, the building area of this commercial complex is only 23,000 m² of over 30,000 m² land area. As for architectural appearance, unadorned black bricks constitute the wall and serve as exterior decorations. The smoky grey wall is only decorated with white brickwork joints. Grey and brick-red buildings alternately arranged show archaic and delicate style and encircle the Presidential Palace in L-shape, creating a commercial building complex of cultural flavor.

总统府是南京民国建筑风貌的集中地，依托于总统府的"南京1912"，体现的也是民国建筑的精神。17幢建筑，其中5幢是原有的民国建筑，最高的只有3层，大多数建筑是两层楼甚至是平房。为了维持风貌，也便于市民休闲观光，占地3万多平方米的一期工程的建筑面积仅为23 000平方米。在建筑外观上，毫无修饰与浮华的青砖既是墙体，又是外部装饰，烟灰色的墙面上，勾勒了白色的砖缝，除此之外再无任何修饰。这片青灰色与砖红色相间的建筑群，风格古朴精巧，错落有致地呈L形环绕"总统府"，成为以民国文化为建筑特点的商业建筑群。

Modern Imitated Commercial Streets of Ancient Style

商业业态与运营
Commercial Activities & Operation

Major Commercial Activities 主要商业业态

Currently, there are more than 50 branding businesses covering leisure, recreation and catering.
Chinese restaurants — Xiao Nan Guo Restaurant, Nihero Cantonese Cuisine, Hou Garden Restaurant, South Beauty and so on.
Western/Specialty restaurants — Restaurant O'Grace, Bellagio, Jiubai Pot and so on.
Bars — Scarlet, Touch Bar, Seven Club and so on.
Tea/Coffee — Starbucks, Costa Coffee, Tea Station and so on.
Leisure — Golden Ladies, David's Camp Men's SPA & Skin Care Center, Xijin Foot Massage and so on.
Others — KFC, Korean Cuisine, Hong Ding Doulao.

街区目前拥有50多家风格各异的品牌商家，覆盖休闲、娱乐、餐饮。
中餐——小南国、银牌自助美食中心、粤鸿和、厚园、俏江南、芳满庭等。
西餐/特色餐饮——王品台塑牛排、逸叶源、泰煌、鹿港小镇、蓝枪鱼、九佰锅、大渔等。
酒吧——百度、乱世佳人、座吧、苏荷、玛索、TOUCH BAR、往事、Seven Club、乌克兰风情吧等。
茶/咖啡——星巴克、Costa Coffee、茶客老站等。
休闲——金夫人婚纱摄影、戴维营专业男子SPA护肤中心、西津足道等。
其他——KFC、韩国料理、红鼎豆捞等。

Operation Measure 运营措施

Operation Mode for Leasing Only
The owner and operator are separated for a complete market-oriented operation. The high-end commercial activities develop the market and the Presidential Palace brings passenger flow. The government also offers support.

During the brand selection process, the business mode should fit the entire project, but the key is the brand awareness, experience and financial strength. The leasing process is the process to subdivide the consumer groups.

The operation mode not only ensures and improves the management level, but also offers a unified social image to the consumers. Since all retails in the block run their shops independently, they can fully display their unique brand image and operation style.

经营方式——只租不卖，管理者与经营者相分离，完全市场化运营。众多高端活动培育市场，"总统府"旅游带动人流，政府作为形象窗口工程提供支持。

在对商家的筛选过程中，除了商家的经营业态要与"南京1912"的定位相吻合以外，最关键的还看品牌、从业经验和资金实力等，招商的过程是细化消费人群的过程。

这种模式可以保证和提高管理水平，更可使街区以一个统一的社会形象面对消费者，同时由于街区内的各零售商分别经营自己的产品，可以充分展示自己独特的品牌形象和经营风格。

品牌商铺展示
Brand Shops

Food & Beverage 餐饮类

Tea Station

In the block, there are several red houses showing the homesickness of the Taiwanese. The houses' owner is the founder of the pearl milk tea and owns the famous tea houses, Onion and Tianming.

茶客老站

  茶客老站是南京1912街区里为数不多的民国红房子,承载着来自台湾的乡愁情结。店主是珍珠奶茶的创始人,在南京小有名气的洋葱和天茗茶楼也是店主的产业。

Dolar Shop

Dolar Shop is a popular Macao hot pot brand managed by Nanjing Catering Management Limited Company. Its product and idea totally subvert the traditional hot pot concept, whilst it inherits the traditional hot pot's essence, combines the tradition with contemporary and incorporates merits of Chinese and Western foods. Its shops with Western décor show the most stylish and innovative dishes. It also has live cooking show and competitive price with exquisite foods and kitchen wares.

典尚豆捞坊

典尚豆捞坊是江苏南京餐饮管理有限公司旗下经营的澳门时尚火锅品牌。它从产品到理念一方面完全颠覆了传统火锅的概念，另一方面又整合了传统火锅的精华，融合了现代与传统、中餐与西餐的优点。它以西式格调装饰餐厅，以美食、美器演绎出火锅的时尚，创新的码盘、现场手工的制作及其优越的性价比，吸引着以时尚都市为荣的多层次消费人群。

Oysters Theme Club

Oyster is the specialty of this restaurant which dedicates to offering innovative design to make the traditional dining a leisurely, stylish, standard and streamlined chain catering industry of the Chinese themselves.

囧吉香

囧吉香主营以"蚝"为特色的创意时尚菜肴。它致力于时尚休闲餐饮品牌的创意设计，把传统餐饮打造得休闲化、时尚化、标准化、流程化，做中国人自己的时尚连锁餐饮。

Hou Garden Restaurant

As a tiny house of the Republic of China, this restaurant features a small fish tank on each dining table and the wall at the staircase of the corner is decorated with various old photos of the city. The word "Hou" means the profound culture of the space. The restaurant mainly caters for Nanjing cuisines, and its archaic décor is very unique.

厚园

　　厚园建筑为民国时期的小楼，每个桌上都有一个小鱼缸，楼梯转角处的墙面上挂满了各种南京的老照片，历史气息浓厚，"厚"指的是餐厅深厚的文化。厚园主营南京菜，古色古香的装修风格颇有特色。

Bellagio

With its name originated from a piece of old song as well as a place, this restaurant has retained the old red bricks of the time of the Republic of China and shows historical sense when completed by its signboard. However, its interior décor is fairly stylish. The food from Taiwan is fantastic. The specialty is ice shaving that you would never have enough.

鹿港小镇

鹿港小镇这个店名源自一首老歌的名字，也是一个地方的名字。建筑保留了民国时期的旧红砖，配上招牌，颇有老店感觉。里面的装修却相当时尚，来自台湾的风味菜肴让人胃口大开，主打菜品是冰沙，令人百吃不厌。

Nihero Cantonese Cuisine

This restaurant leads a trend of eating top Cantonese cuisines in Nanjing. The façade in black and white is adorned with private collected old photos of the Cantonese female boss. There also shows silent film with black-and-white images on a giant screen. The improved dishes are quite pleasant to the Nanjing people.

粤鸿和

粤鸿和在南京成功掀起"燕鲍翅风尚",建筑立面的主色调是黑白两色,墙上挂的老照片都是广东女老板的私人藏品,大屏幕播放着无声黑白老电影。菜式经过改良,特别适合南京人的口味,"金陵第一鸽"和"海鲜粉丝煲"已成为脍炙人口的招牌菜。

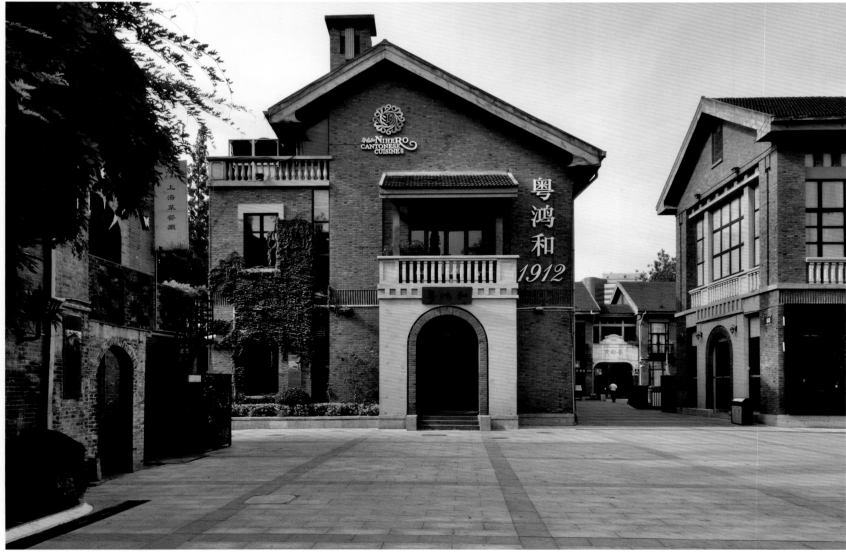

Jiayue

It was praised as "the best Japanese restaurant" by Nanjing citizens. Its main products include silver pout, udon noodles and so on. Its dishes are top-class and exquisite, also expensive.

嘉月

嘉月被南京人誉为"最棒的日式料理店",菜品主要有鲜嫩多汁的银鳕鱼、富有韧劲的乌冬面等,高档精致,价格不菲。

Taihuang

Its main products are Thai-style cuisine and seafood hot pot. All dishes are made with natural plants instead of monosodium glutamate.

泰煌

泰煌以泰式料理、海鲜火锅为主,所有食品不加味精、酱油,调料全用泰国的自然植物代替。

Blue Marlin

Blue Marlin features European-style dishes and selects classic cuisines from Asia and America.

蓝枪鱼

蓝枪鱼以欧陆式菜肴为主,并精选了亚洲、美洲的经典饮食,目前已拥有5家连锁店。

Entertainment 娱乐类

Bangbangtang KTV
This KTV has reserved the original features of the buildings of the Republic of China. Some post modern elements are also added to show its modern style.

棒棒堂量贩式KTV

棒棒堂量贩式KTV保留了民国建筑的原有特色，加入了一些后现代风格的元素增加店面的现代感。

Mazzo-V-club

The word "Mazzo" in French means night club chased after by numerous men and women. The shop owner, a legend man in Chinese bar industry, is also a master in bar décor, music and business operation. In this club, all designs including sopraporta, crystal droplight, Alligator sofa and so on are bespoke to his imagination.

Mazzo-V-club

"Mazzo",法语中意为"夜店",是男人的最爱,也受女人的追捧。店主解刚是中国酒吧界传奇人物,从装潢设计、音乐到经营样样精通,酒吧大到门头,小到水晶灯、鳄鱼皮沙发,都是根据他的想法定做的。

RiCHY International Entertainment

Themed by Baroque style and integrated with the elements of the Middle East in the 20th century, RiCHY International Entertainment makes a breakthrough in music style, stage performance and interactive entertainment. Meanwhile, it also gathers entertainers regularly and is the first choice for fashionistas.

瑞奇国际娱乐

瑞奇国际娱乐以巴洛克风格为主题，融入20世纪中东元素，在音乐风格、舞台演绎、娱乐互动方面进行突破性创新，更有娱乐人物定期汇聚，是时尚人士的首选之地。

Glimpses of Paradise

Glimpses of Paradise, founded in 2004, is a famous brand in hairdressing field of Nanjing. It has won a very high praise in Nanjing fashion field and high-end customers.

天籁艺境

天籁艺境于2004年创办,是南京美发行业的知名品牌,在南京时尚界和高端客户中拥有极高的知名度。

Others 其他

Golden Ladies

Golden Ladies, leading the trend of China bride photo industry, establishes its flagship store of eastern China in Nanjing. It is also the only wedding photo studio which can take pictures in Nanjing 1912 .

金夫人婚纱摄影

中国婚纱摄影界领头羊"金夫人"在南京建立华东地区旗舰店,它是唯一一家可以在南京1912街区取景拍摄的婚纱摄影影楼。陈慧琳2005年南京演唱会指定由金夫人做整体造型。

1912 Hefei
合肥1912

Street Background & Market Positioning
街区背景与定位

History 历史承袭

Dashu Mountain is located in Hefei, Anhui. It was formed by vulcanian eruption and is higher at southeast and lower at northwest in oval shape. Now the Shushan District is located at the southwest part of Hefei with Jinzhai Road and Baohe District to its east, Huancheng West Road, Nanfei River and Luyang District to its north and Feixi County to its west and south. It is a burgeoning district with development vitality and high urbanization level.

大蜀山位于安徽省合肥市境内，系大别山余脉，山势东南高、西北低，呈椭圆形，由火山喷发而成。据《庐州府志》记载，"有蜀僧于此结庐，偶思乡水以锡卓地，泉汩汩而出，尝之有瞿塘峡味，因名为蜀井"，蜀山以此而出名。

现在的蜀山区位于合肥市西南部，东以金寨路与包河区为界，北与环城西路、南淝河、庐阳区为邻，西、南两面与肥西县相接。区域内坐落着国家级合肥高新技术开发区、合肥经济技术开发区、合肥政务文化新区和正在建设中的中国（合肥）科学城，是一个以科技文化教育为先导、高新技术产业为主导、高效社会服务为引导，极具发展活力、城市化水平最高的新兴城区。

Location 区位特征

1912 Hefei is located at Shushan District, one of the four big districts in Hefei. It is between the first ring and second ring of the city and is adjacent to Huangshan Road, an important city landscape road. It is surrounded by up-scale residences, scientific research institutions and colleges.

合肥1912位于合肥四大主城区之一的蜀山区，地处城市一环与二环之间，紧临合肥重要城市景观大道——黄山路。项目西望大蜀山，南接合肥政务文化新区，两大国家级高新区、经济开发区分踞其西侧和南侧，周边高尚住宅、科研机构、高等院校林立。

Market Positioning 市场定位

It is positioned as the first large scale commercial block featuring culture creativity and gathering bars, leisure, entertainment, restaurants, residences, tourism and shopping in Hefei.

项目为合肥首个以文化创意为命题，集酒吧、休闲、娱乐、餐饮、居住、旅游、购物为一体的大型特色商业街区。

规划设计特色 / Planning & Design Features

Street Planning 街区规划

Form: adopting BLOCK planning design to add street area and drive interior stream of people, this project is divided into three parcels, among which A and B are for business and C is for residence.

Commercial activities: restaurant, leisure and entertainment, cinema and hotel are introduced to bring more night consumption.

Commercial activities distribution: anchor stores are distributed at supporting point of the parcels to drive the stream of people of sub-anchor stores. Sub-anchor stores are distributed intensively to form a layout of different industry groups. The whole area generally forms three commercial streets.

形态：采用BLOCK街区规划设计，增加沿街面积，带动内部人流。项目分为A、B、C 3个地块，A、B两地块打造商业项目，C地块打造住宅项目。其中商业建筑面积为80 000平方米，住宅面积为180 000平方米。

业态：除了引进餐饮、休闲、娱乐业态外，还引入了原先没有的影院以及精品酒店等。通过影院的介入以及酒吧带来的夜间消费，让餐饮和休闲商家在夜间也能获得收益。

业态分布：主力店分布在地块支点，带动次主力店人流；次主力店集中分布，形成各业态集群分布格局。整体形成3条商业街：马鞍山南路复式铺约5 000平方米，主要面积区间为70～100平方米；步行街分层独立铺共5 500平方米，主要面积区间为35～50平方米；规划路复式铺共6 500平方米，主要面积区间为80～100平方米。

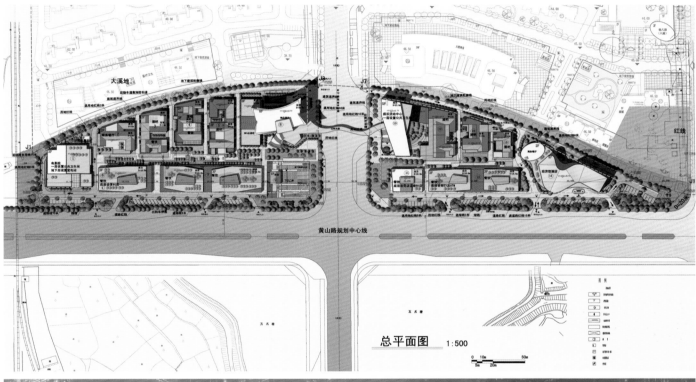

总平面图 1:500

Street Design Features 街区设计特色

Fashionable Block + Hui-style Elements

Hui-style culture is brought in to integrate modernity with tradition. A themed commercial street as the axis links up restaurant area and bar area along Huangshan Road with cinema, hotel interspersed. Several themed courtyards and squares are formed between streets and buildings to create an outdoor commercial place full of atmosphere. Sunken yard and old theatrical stage at the west parcel and Huangmei Theater at the east parcel form a visual center and space-time node. History and future, culture and business meet here harmoniously.

Combination of Universality and Customization

The design of single commercial building follows the way of combing universality and customization. Meanwhile balconies, terraces and yards are designed to create social space to improve communications between people and enhance business value.

Tradition and Modernity Interactively Presented

There is a sunken center square which leads into the old theatrical stage. An extremely strong visual impact was formed between the old theatrical stage and hotels. U-shaped glass, corrugated sheet steels, ventilation louvers and so on are elaborately used to realize a delicate impression experience of cultural atmosphere.

Comparison and Combination of Different Colors

Buildings along the Huangshan Road use aluminum plates of warm colors of similar hue to avoid tedium. These colors contrast with that of Dashu Mountain not far away. Meanwhile, buildings of Southern Anhui style are arranged among commercial buildings, thus grey and white contrast with other colors arranged to achieve a distinctive visual effect.

时尚街区+徽派元素

设计引入徽派文化，将现代和传统进行融合。在空间处理上，以一条主题商业内街为主轴，将沿黄山路的高档餐饮区和内部的情景合院式酒吧区串联，电影院、精品酒店、娱乐中心、经济型酒店穿插其中，既形成较为醒目的地标，亦成为情景商业的绚丽背景。内街和建筑、建筑和城市界面之间形成多个主题庭院和广场，结合鱼骨状的传统街巷空间，营造出极具气氛的室外商业场所。西侧地块的下沉庭院、古戏楼以及东侧地块的黄梅戏院，形成视觉中心和时空节点。历史与未来、文化与商业在此邂逅，和谐而统一。

普适性原则和量身定做相结合

商业建筑单体设计遵循普适性原则和量身定做相结合的方式，既满足了业主对使用功能灵活可变的要求，又满足了业态配比，保证了重点商铺的发展。同时，设计各具情趣的退台、露台、庭院，实现了交往空间的营造和关怀，促进了人与人的交往，提升了商业价值。

传统与现代交互呈现

项目的中心广场有一个下沉式的中心广场，通过下沉式的中心广场引入古戏台，古戏台在酒店之间，形成非常强烈的视觉冲击感。项目力求创新，非标节点的U形玻璃、波形钢板、金属瓦屋面、隐蔽式通风百叶、点支玻璃肋幕墙等，经过反复推敲和深化，以期实现一个有着文化气氛的、精细的感观体验。

不同色彩的对比组合

在黄山路沿线的建筑中，设计师采用暖色调且色相接近的铝单板，并把它们并置在一个建筑平面中，显得含蓄和谐，避免了同一色相下的单调与沉闷，与不远处生机盎然的大蜀山形成色调上的冷暖对比，同时又相互映衬，拉近了消费者与街区的距离。皖南建筑中的粉墙黛瓦被布置于商业建筑中，其灰白色和彩色形成了对比，取得了鲜明的视觉效果。

Major Commercial Activities 主要商业业态

Commercial Activities 商业业态

1912 Hefei is the only one cultural commercial street organically combining Hui-style architecture and modern architecture in Anhui. It has rich commercial activities including D-MAX cinema, restaurant street, bar street, hotel, KTV, SPA and so on.

A Area —— China Film Cinema, Return 97 Bar, Mangoo, Yeyan Club, Club TNT, Blue Marlin, Yakiniku Master, Shiga Hous, Qiqi Hotpot, Lavazza Italian Café, Shanghai Pudong Development Bank, Starbucks, Fenglin International Hotel, and so on.

B Area —— Wanjiale Cartoon Happy World, Boli Liugang SPA, Yundingzhidian Snooker Club, Basjoo Leaf Tea Art Club, Tianxianpei Stage, South Beauty, Huidingge Hotel, Rococo Dessert, 1862 Restaurant, Red Diamond Bar, Bacchus, Tang Club, Maksim, and so on.

合肥1912是安徽省唯一一个徽派建筑与现代建筑有机组合的文化特色商业街区，拥有丰富的商业业态，包括D-MAX电影院、餐饮街、酒吧街、酒店、KTV、黄梅戏院、古戏楼、SPA会馆等，是集吃、喝、玩、乐、观光为一体的高端文化、时尚休闲街区。主要商铺如下。

A区——中影国际影城、回归97酒吧、芒可、夜艳会所、潮人会所、尚坐Club、蓝枪鱼、九品风山大宝、烹大师烧肉达人、春娇与志明、贺滋屋、齐齐火锅、LAVAZZA意式生活馆、浦发银行、至尊烟酒、尿尿小童炸鱼薯条、星巴克、枫林国际酒店、开放式戏院等。

B区——玩家乐动漫欢乐世界、玻璃流光SPA、艾莎足道、云顶之巅桌球会、芭蕉叶茶艺会馆、天仙配大舞台、俏江南、祥记燕翅鲍、徽鼎阁、洛可可甜品、禾禾苑日式料理、1862休闲餐厅、红钻演艺酒吧、Bacchus、唐Club、皇马会所、马克西姆、欧堡万国酒庄、皇庭国际娱乐会所、艾微精品主题酒店等。

品牌商铺展示
Brand Shops

Food & Beverage 餐饮类

South Beauty

South Beauty, founded in 2000, is the most promising international restaurant service group in China. South Beauty brand restaurant integrating the Oriental and Western culture is boutique restaurant of unique Sichuan style.

俏江南

俏江南集团创始于2000年，总部位于北京，旗下包括俏江南品牌餐厅、兰会所和SUBU三大高端品牌，是中国最具发展潜力的国际餐饮服务集团。俏江南品牌餐厅是集东西方文化为一体、具有独特韵味的四川精品餐厅，特色菜肴有江石滚肥牛、麻酱油麦菜、金牌水煮鱼、口水鸡、晾衣白肉、金牌过桥排骨、毛血旺、俏江南手撕鸡、浓汤娃娃菜、泡椒凤爪、文房四宝、水煮鲶鱼、鱼唇汤等。

Yakiniku Master

Yakiniku Master was founded by Eric Tsang, a famous Hong Kong film star, and an experienced group. It has exquisite cooking skills and a large scale central kitchen research and development base which establish a solid foundation for its rapid development.

烹大师

烹大师炭火烧肉由香港著名影星曾志伟及一个具备丰富餐饮管理经验的团队共同创立,是烹大师餐饮集团继烹大师火锅、烹大师涮涮锅之后的又一新作。精湛的烹饪技术、大型的中央厨房研发基地为其快速发展奠定了坚实的基础,多项菜色为其专业研发秘制而成。

Starbucks

Starbucks not only offers coffee, but also conveys a unique style to customers through the carrier of coffee. It creates an environmental culture to affect customers and to provide customers with good interactive experience.

星巴克

星巴克的产品不仅是咖啡,而是通过咖啡这种载体,把一种独特的格调传送给顾客。咖啡的消费很大程度上是一种文化层次上的消费,文化沟通需要的就是咖啡店所营造的环境文化能够感染顾客,让顾客享受并形成良好的互动体验。

Entertainment 娱乐类

Baccbus

Baccbus is a themed bar offering authentic European and American food, and famous German beer. The European and American style of Baccbus is rare in Hefei.

巴克斯

巴克斯酒吧是以经营正宗欧美西餐、知名德国啤酒为主的主题餐吧。巴克斯的欧美风情是合肥极为少见的风格,素有"老外聚集地"之称。

Club TNT

It has unique hip-hop fashionable music, antique interior design with simple lines, high-quality sound equipment and brand new lighting design.

潮人会所

此会所拥有独一无二的时尚潮流音乐,简洁的线条、复古式的室内设计、优质的音响设备、全新的灯光舞美设计将潮人会所打造成一个全新的聚会殿堂。

Return 97 Bar

It is a high-class bar subordinating to Wuhan Return Entertainment Management Co., Ltd. which was founded in 1997.

回归97酒吧

回归97酒吧是中国武汉回归娱乐管理有限公司旗下的高端酒吧,创立于1997年,是一家集餐饮娱乐为一体的大型娱乐管理集团企业。经营范围主要包括迪吧、演艺厅、KTV等。

China Film Cinema

China Film Cinema Group was founded in 1999. It is the biggest and most completed film industry group in China.

中影·国际影城

中国·国际影城是中国电影集团旗下的分公司。中国电影集团公司（简称"中影集团"）于1999年成立，由原中国电影公司、北京电影制片厂、中国儿童电影制片厂、中国电影合作制片公司、中国电影器材公司等8家单位组成，是中国目前最大、最全的电影产业集团，产业链包括制片、制作、宣传营销、发行放映、影院投资、海外营销等6大环节。

Accommodations 住宿类

Anyway Boutique Hotel

It is the first high-class boutique hotel in Hefei. It was designed elaborately by famous designer who used fashionable and tasteful trendy elements for reference. There are 14 kinds of rooms of different themes. All the rooms are equipped with completed facilities including TV, central air-condition, free Internet, and so on.

艾唯精品主题酒店

艾唯精品主题酒店是目前合肥第一家高档精品主题酒店，由名家精心设计，借鉴了兼具时尚和情调的潮流元素，环境舒适。酒店有14种主题风格迥异的房间，每种房型都配备42寸超大豪华液晶电视、中央空调、免费光纤上网、24小时安全监控等系列服务设施；浴室设备、网络配置、床上用品等方面，完全参照星级标准。

Chu River & Han Street, Wuhan
武汉楚河汉街

街区背景与定位 Street Background & Market Positioning

History 历史承袭

Wuhan Central Culture Tourist Area was invested and developed by Wanda Group uniting with top design companies. Chu River & Han Street, the first phase project of Wuhan Central Culture Tourist Area, is the start-up project of Wuhan Dadong Lake ecological water network construction project and the core project marking the 100th anniversary of the Revolution of 1911.

武汉中央文化区项目由万达集团投资开发，项目整体规划由万达商业规划院牵头，联合国内外业内顶尖设计公司参与完成设计。武汉中央文化区一期项目楚河汉街是武汉市大东湖生态水网构建工程的启动工程、纪念辛亥革命100周年的核心项目。2011年9月楚河汉街开业，在国庆假期吸引了超过200万人次的客流，成为全国假期人流排名前三的热点区域。

Location 区位特征

Chu River & Han Street is located at the core area of Wuhan, from Donghu Road at the east to Yanxia Road at the west. It is adjacent to Gongzheng Road Bailu Street at the south and Chuhe Road at the north. Chu River is 2.2 kilometers long connecting East Lake and Sha Lake. Han Street lying in the south of Chu River is 1.5 kilometers long.

楚河汉街位于武汉市核心地段，东临东湖路，西至烟霞路，南接公正路白鹭街，北依楚河路，项目总建筑面积21万平方米。楚河全长2 200米，连通东湖和沙湖，汉街商业步行街偎依在楚河南岸，全长1 500米。楚河汉街引入众多国际知名品牌，商业繁荣；街内以民国时期建筑风格为主，打造众多主题广场，文化氛围浓厚。

Market Positioning 市场定位

Chu River & Han Street is a "kingdom in a city" which centralizes cultural tourism and contains commerce, food, leisure and entertainment. It is a brand new landmark of recreation, fashion and art. It is trying to be the most fashionable and charming waterside commercial street in the country.

由万达集团携手世界知名建筑设计师设计的汉街位于楚河南岸，总长1 500米，是一座以文化旅游为核心，兼具商业、美食、休闲、娱乐为一体的"城中王国"，也是武汉全新的娱乐湾、潮流港、不夜城、娱乐餐饮地标、流行时尚地标、人文艺术地标，力图打造成为国内档次最高、最时尚、最具魅力的沿水商业街区。

规划设计特色 / Planning & Design Features

Street Planning 街区规划

After studying the history, commercial layout and water system of Wuhan City, the architecture design of Han Street was made. It focuses on ecology, relies on culture and is based on the style of Republic of China. The designs of buildings at Han Street are divided into three parts to realize the integrity of style.

The east part — buildings of Republic of China combining Chinese and Western elements as basic tone, this part chooses continuous arcade building to build a modern commercial street with traditional building style. By smart collocation of colors, sizes, traditional and modern elements, the design achieves a concept of "from history to future".

The middle part — absorbing the advantages of buildings in the Republic of China and concession, plain brick wall and modern glass curtain wall are combined to present the continuity of historical culture and integration with modern culture. Near to Wanda Plaza, the style therefore is more modern and harmonious to Wanda Plaza.

The west part — partially breaking through Chinese and Western mixed style of the Republic of China and putting more emphasis on innovation, the form of gables is more changeable and more creative and the form of roofs expresses the improvement of buildings in the Republic of China by the modern design idea.

通过对武汉城市历史、商业布局、水网工程进行深入研究和理解后，汉街建筑设计确定为以生态为原点、以文化为依托、以民国风格为基础的金字塔结构。同时，设计师根据项目周边的人文景观等资源，对汉街的建筑风格分段设计，以实现风格完整、特色突出的建筑形态。具体分段如下。

汉街东段：以中西合璧的民国建筑为基调，选取连续的骑楼及个性强烈、变化丰富的山墙面等传统建筑元素，打造具传统建筑风格的现代商业街；设计利用色彩、体量、传统与现代元素的巧妙搭配，实现从历史走向未来的设计立意。

汉街中段：吸收武汉民国时期建筑与租界建筑的优点，强调兼收并蓄；清水砖墙与现代玻璃幕墙元素的结合体现武汉历史文脉的延续与融合；临水建筑的立面构图吸收民国时期建筑的优点，展现独特气质；紧贴汉街万达广场街区，风格形式更加现代，与大体量的汉街万达广场协调统一。

汉街西段：建筑局部突破中西混合的民国风格，更加强调"推陈出新"，使得山墙形式更加多变和富有创造力。屋顶形式也表现出当代的设计思想对民国建筑的改进；大体量的现代建筑通过形体的叠加和弱化，与周边三层的街道尺度达到和谐统一。

Street Design Features 街区设计特色

The main body of Han Street mainly adopts the building style of Republic of China. When walking on the street, visitors will feel like going back to old times. Meanwhile, modern buildings and European style buildings are alternated among the buildings of Republic of China style to achieve a perfect mixture of the tradition and the modernity. There are five theme plazas which add cultural ambience to Han Street. Riverside landscape belts are built on both sides of Chu River and luxury yachts sailing from Sha Lake to Chu River for visitors to enjoy the scene of the water.

In addition, the design of lighting of night scene is also very appropriate. The lighting show basing on the five theme squares supplies various kinds of technique to highlight the 3D sense of buildings and exquisite details. And the lighting of the riverside is divided into several layers from waterside platform to the railing and then to the façade of buildings. The light color is gradually transformed from quiet and graceful blue white to comfortable and pleasing warm white.

汉街主体采用民国建筑风格，处处是红灰相间的清水砖墙、精致的砖砌线脚、乌漆大门、铜制门环、石库门头、青砖小道、老旧的木漆窗户，置身其中，仿佛时光倒流。同时，汉街中又将时尚的现代建筑和欧式建筑穿插在民国风格建筑中，实现传统与现代的完美融合。另外，汉街内有屈原广场、昭君广场、知音广场、药圣广场、太极广场五大主题广场，广场均由几部分组成，营造出汉街的文化气息。楚河作为重要的景观，两岸设有滨河景观带，成为武汉市民惬意的亲水平台，河内有豪华游艇从沙湖驶向楚河，供游客于河中观赏沿岸景观。

汉街夜景也是汉街设计之一。汉街夜景的照明设计十分妥当，灯光秀主要以五大广场为依托，采用了多种照明手法以及先进的照明技术，以大范围的泛光照明来渲染建筑立面，显示出立体感，突出细部的精致，体现建筑自身独有的构造，营造矜持而低调的奢华，打造出整个汉街的底色。汉街滨水河岸的照明分几个层次，从亲水平台到白玉栏杆，再到建筑立面，形成不同的光带，光色由宁静幽雅的蓝白色光过渡到舒适宜人的暖白光。

Major Commercial Activities 主要商业业态

The Han Street gathers over 200 first-rate merchants and shops at home and abroad, including shopping, restaurant, culture, leisure and entertainment. As for shopping, there are international fashion brands including ZARA, H&M, C&A, GAP, MUJI, M&S, UNIQLO, UR and so on, firstly realizing the grand occasion of gathering the global ten top fast fashion brands in the world. As for restaurant, there are Starbucks, Häagen-Dazs, McDonald, Tsui Wah Restaurant, Cheers Palace, Sunshine Kitchen and so on. As for culture, there are world famous Madame Tussaud's, Disney flagship store, book store and art gallery.

汉街集合了200多个国内外一流商家，包括购物、餐饮、文化、休闲、娱乐等，商业十分繁荣。在购物方面，汉街聚齐了包括ZARA、H&M、C&A、GAP、MUJI、M&S、UNIQLO、UR等全球时尚流行品牌，首次实现世界十大快时尚品牌毗邻而居的盛况。在餐饮方面，汉街吸引了诸如星巴克、哈根达斯、麦当劳、翠华餐厅、迎蕾皇宫、汤城小厨等诸多品牌餐饮店入驻。在文化上，世界著名的杜莎夫人蜡像馆、拥有全线产品的迪士尼旗舰店、文化书城以及中国著名的正刚艺术画廊等文化品牌纷纷落户汉街。另外，汉街还成为国内外知名品牌开拓武汉乃至中部市场的登陆点，如鹿港小镇、青花元年、周大福集合店、90plus、论道生活馆等。

Clothes 服饰类

M&S

M&S (Marks & Spence) from Britain is a representative chain store with a-hundred-year history. It mainly offers clothes and food and is one of the top ten fast fashion brands of the world.

M&S

　　M&S（Marks & Spencer）源自英国，是英国拥有百年历史的代表性的连锁商店之一，以销售服饰和食品为主，是世界十大快时尚品牌之一。

Food & Beverage 餐饮类

Cheers Palace

Cheers Palace is a subordinate brand of Hong Kong Tao Heung Group, the largest listed Chinese style restaurant company in Asia featuring professional banquet operation. The Cheers Palace at Han Street was opened in September 2012. Its site design is leaded by European style and the wedding party arrangement integrates Chinese and Western elements. Its dishes are mainly Hong Kong-style Cantonese cuisines which are healthy, light and of low calorie.

迎囍皇宫

"迎囍皇宫"是亚洲最大的上市中式餐饮集团香港稻香集团旗下的品牌之一，主打专业宴会业务，于2010年、2011年连续两年荣获"中国广州饮食天王——婚宴天王"的称号。位于汉街的迎囍皇宫于2012年9月开业，场地设计以欧陆风格主导，引入中西合璧的婚宴场地布置，华丽设计配以LED大型屏幕及大剧院式的舞台，可筵开60席的超大场地，彰显皇家盛宴气派。迎囍皇宫菜式以正宗港式粤菜为主，健康、清淡、热量低。

90 PLUS

90 PLUS is a chain wine store which is a professional wine platform brand subordinated to Beijing Capital Wine Co., Ltd. 90 PLUS is not a single brand; rather, it is a brand composed by superior quality wine selected by world wine critics. 90 PLUS offers over 500 kinds of drinks from all over the world and its staffs in the shop are all trained systematically and professionally.

90 PLUS

90 PLUS连锁酒屋是北京紫禁红贸易有限公司旗下的专业葡萄酒平台品牌。90 PLUS不是某一支葡萄酒产品的品牌，而是经全球酒评师精选的优质葡萄酒（即百分制评测得90分以上者）的一个集成品牌。90 PLUS连锁酒屋经营来自世界各地的500多种酒水类产品，店内员工都接受过系统的酒水类课程培训，不仅可以向顾客提供优质的酒水类产品，还可以为顾客提供咨询、向导服务，帮助消费者做出选择。

Hankou Jingwu

Hankou Jingwu mainly sells spiced ducks including duck's necks, duck's gizzards, duck's tongues and duck's feet. Modern processing technology and traditional workmanship are combined to make delicious products which are very popular in customers.

汉口精武

汉口精武主营卤鸭食品，包括鸭脖、鸭胗、鸭舌、鸭掌等，其食品采用现代食品加工技术、融合传统工艺、辅以几十种名贵香料、科学加工精制而成，成品微辣，咸甜适中，肉质细嫩，高蛋白，低脂肪，回味持久，深受顾客喜爱。

Photography 摄影类

AVIVA Photograph

AVIVA Photograph was founded in 2008 in Wuhan. Its headquarters is at the foot of Guishan Mountain. "AVIVA" is a Hebrew word meaning beautiful tomorrow and is extended to "hope". It devotes to China wedding field and provides customers with various styles of photographs, customized wedding dresses, wedding following shots and so on.

薇拉摄影

薇拉摄影于2008年创立于武汉，总部坐落在龟山脚下。"AVIVA"原是希伯来词语，意为"美好的明天"，引申意为"希望"，寓意将美好的祝愿送给所有的朋友。薇拉摄影致力于中国婚嫁行业，提供多种类、多风格婚纱摄影、婚纱礼服定制、婚礼化妆造型定制、婚礼跟拍等服务。

文化设施 / Cultural Facilities

Han Street Big Stage 汉街大戏台

The Han Street Big Stage is located at the east of Eternal Friend Plaza. It is used for public show for free. The stage is built in antique wooden structure and its backdrop applies the most advanced LED screen in the country. On holidays, there will have traditional Chinese opera shows which enrich the spare-time life of citizens.

汉街大戏台位于知音广场东侧，用于群众免费演出。戏台是仿古的木式结构建筑，背景采用国内最先进的LED屏幕。节假日时，汉街大戏台上会上演中国传统戏曲艺术，丰富武汉群众的业余文化生活。

Five Plazas 五大广场

There are five plazas named after celebrities of Hubei at the Han Street. They are Quyuan Plaza, Zhaojun Plaza, Eternal Friend Plaza, and Medicine Master Plaza at east entrance and Tai Chi Plaza at west entrance. Each plaza has sculptures and landscape according to its theme.

汉街设有5座以湖北地区历史名人命名的广场，分别是位于东入口的屈原广场、昭君广场、知音广场、药圣广场和位于西入口的太极广场，每个广场内都按照广场的主题设有一座名人雕塑及相关景观，如屈原广场中设有楚文化地面石雕和音乐喷泉。

1. 屈原广场

2. 昭君广场

3. 知音广场

4. 药圣广场

5. 太极广场

引导指示系统
Guidance & Sign System

Chenghuang Temple Commercial Street, Shanghai
上海城隍庙商业街

街区背景与定位 / Street Background & Market Positioning

History 历史承袭

The Chenghuang Temple was built in Yongle Period of Ming Dynasty, and was expanded for several times during Ming and Qing dynasties, After Daoguang Period of Qing Dynasty, the scale of Chenghuang Temple became smaller and smaller for storm and stress of the society, and finally closed down in 1966. In 1991, Chenghuang Temple Market was turned into a modern large tourism shopping center of traditional features by Shanghai municipal government. In 1994, a series of restoration of Chenghuang Temple was done by the support of the government and Taoists. Now the Chenghuang Temple Commercial Street is a must-go tourism area.

上海城隍庙始建于明代永乐年间，明清两代曾多次扩建，极盛时庙基达49.9亩，约33 000平方米。清代道光以后，因为社会动荡，上海政局不稳，加上新中国成立后受"左"倾思想影响，城隍庙庙基不断缩小，于1966年关闭。1991年上海市政府开始将城隍庙市场改建成为具有传统特色的现代化大型旅游购物中心。1994年城隍庙重新开放，之后，在政府与道教信徒的支持下，城隍庙完成了一系列的修复工程。现在上海城隍庙商业街已成为到上海游玩一个必去的旅游街区。

Location 区位特征

Shanghai Chenghuang Temple Commercial Street is located at both sides of Fangbang Middle Road, Huangpu District. The Chenghuang Temple Street which has over 600-year history has a profound cultural connotation. Till early Republic of China, it had been the political, economic and cultural center of Shanghai. Now it has the developed commerce and rich tourism with "Yuyuan", one of the top three Jiangnan classic gardens of Shanghai in Ming Dynasty in this area.

上海城隍庙商业街位于上海市黄浦区方浜中路左右两侧，东至安仁街，北至福佑路，西达旧校场路。上海城隍庙街区的历史文化底蕴深厚，拥有600多年的历史，直到民国初期，一直是上海的政治、经济和文化中心；商业发达，旅游景点丰富，明代上海三大名园之一的江南古典园林"豫园"位于区内。

Market Positioning 市场定位

The Chenghuang Temple area which has been called "the root of Shanghai" is a peculiar cultural mark of Shanghai. Its market positioning is to make Shanghai Chenghuang Temple Commercial Street into a international tourism spot gathering temples, garden architecture, shops, food and tourism, by digging its profound cultural connotation and showing the strong folk customs of Shanghai.

上海城隍庙地区从元代到民国初年，一直是上海的政治、经济、文化中心，被称为"上海的根"，是上海特有的人文标志和文化名片。它的市场定位是深挖其丰厚的文化底蕴，通过数百年历史文脉的物化展示和上海城市文明的视觉演绎，展现上海浓郁的民俗风情，将上海城隍庙商业街打造成为一个集邑庙、园林、建筑、商铺、美食、旅游等为一体的国际旅游目的地。

规划设计特色
Planning & Design Features

Street Planning 街区规划

The planning of Chenghuang Temple Commercial Street is positioned as a style and feature protection zone giving priority to garden style market. Centralizing historical relic such as Chenghuang Temple and Yuyuan, this area is graded scope of protections and the heights of buildings are controlled to keep the overall form. The original city texture is preserved to present its historical style and feature.

"Integrity" theory of urban design by Alexander was adopted when planning Chenghuang Temple Commercial Street. A square was set in front of the main entrance of Chenghuang Temple. Design strategy of "breaking up the whole into parts" was supplied to avoid the contradiction between traditional city texture and large buildings. The theatrical stage and Chenghuang Temple are linked by a square through inside street. By integrally controlling all the construction process, each individual building is connected to each other, and therefore making sure that the traditional connotation of city space is recovered and memories of the city are inherited.

　　城隍庙商业街的规划定位为以园林商市为主的风貌保护区，以城隍庙、豫园等文物建筑为中心向外辐射状分级划定保护范围和控制要求，并通过控制建筑屋脊高度来控制总体形态，保存原有城市肌理以体现历史风貌。该地段的街道，以前多随水延伸，或填河造路，呈现出蜿蜒自然的形态，规划中保留了这种类似于江南水乡河道或小巷的宜人尺度。

　　城隍庙街区在规划时采取了Alexander城市设计的"整体性"理论。设计者在正对城隍庙正门的位置设置广场，以此为节点加强了基地与城隍庙的联系，并以规划的圆形和考究的影壁、牌坊等传统标志凸显城隍庙与整个区域的地位。项目采用"化整为零"的设计策略，在旧有的城市肌理完整的地区分散大体量建筑，避免传统的城市肌理与大体量建筑的矛盾，并通过内街将广场与方浜中路街角的戏楼产生联系，使戏楼通过广场与城隍庙联系。通过整体性控制所有的建设过程，使一座单体建筑与另一座单体建筑彼此兼顾、链锁式发展，为恢复城市空间的传统意蕴和延续城市记忆的城市空间提供保证。

Street Design Features 街区设计特色

The Chenghuang Temple Commercial Street has a very long history. Buildings in the street mostly imitate Ming and Qing style. There are various kinds of shops and every of them is different from each other. The street keeps Chinese old town style and feature.

Chenghuang Temple is a must-visit scenic spot. Visitors can pray for safeness for their families in the main hall. The Jiuqu Bridge on the Lotus Pond has a unique style. With turns, visitor can enjoy different views as moving on the bridge. And pass the Mid-lake Pavilion is the Yuyuan which is a classic Jiangnan garden. Then at the Yuyuan Old Street, there are many shops of different characteristics, also famous restaurants and teahouses. At night, the Chenghuang Temple Commercial Street is even more charming.

上海城隍庙商业街历史悠久，街中建筑多仿明清风格建造，两旁商店鳞次栉比，商品琳琅满目，各具特色，保持着中国古老的城镇街区风貌。

在城隍庙商业街中游玩，古老沧桑的城隍庙是必去的景点，穿过一座宫殿，于正殿前敬畏地烧柱香，祈求家人平安。荷花池上的九曲桥独具风格，三步一折，移步换景，颇多趣味。荷花池中悠闲自在的金色鲤鱼，不时地将闲适的感觉传给游人。过湖心亭，便到豫园，一座古典的江南园林小巧别致，满园的绿意消除游人的丝丝疲乏。出豫园，到豫园老街闲逛，各家商铺独具特色，琳琅满目的老字号商品，既饱人眼福，又满足游人的购物欲望。还有古色古香的绿波廊酒楼中地道的上海菜让游人一饱口福，豫园西苑中"禅武神韵"的中国功夫表演让人流连忘返。入夜，仿明清建筑的飞檐翘角被节能光灯镀上美丽的金边，四周幽静、安逸。九曲桥边，荷花池畔，三三两两的游人在此小憩，低声交谈，城隍庙商业街显得越发优雅迷人。

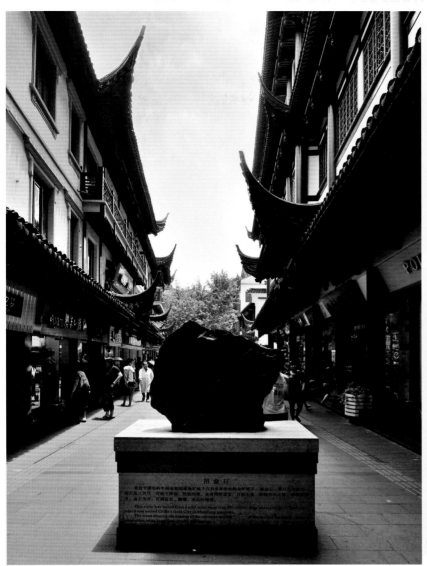

Modern Imitated Commercial Streets of Ancient Style

Modern Riverside Scene at Qingming Festival
现代清明上河图

Major Commercial Activities 主要商业业态

The commerce of Shanghai Chenghuang Temple Commercial Street is mainly distributed in Yuyuan Tourist Mart which originates from the old Chenghuang Temple over 150 years ago. There are various kinds of industries in Yuyuan Tourist Mart.
Gold and jewelry: Shanghai Old Temple Gold, Shanghai First Asia and Laofengxiang.
Food & beverage: Shanghai Classical Hotel, Lvbolang Restaurant, Songyunlou Restaurant, Laosongsheng Restaurant, Old Chenghuang Temple Snack Square, Chunfeng Songyuelou Vegetarian Restaurant and so on.
Pharmacy: Tonghanchuntang.
Department store: Tianyu Department Store, Yuyuan Old Street, Yuyuan Fashion Street.
Food: Spiced Bean Store, Pear Syrup Store and so on.

上海城隍庙商业街的商业主要分布在豫园商城之中，豫园商城源于150多年前清代同治年间的老城隍庙。凝聚百年经典的豫园商城内分布着黄金珠宝、餐饮、医药、百货、食品等诸多行业。具体的行业商铺、街区分布如下。
黄金珠宝业：上海老庙黄金、上海亚一金店、老凤祥。
餐　　饮：上海老饭店、绿波廊酒楼、松运楼酒家、老松盛、老城隍庙小吃广场、春风松月楼素菜馆等。
医　药　业：童涵春堂。
百　货　业：天裕百货、豫园老街、豫园时尚街。
食　品　业：五香豆商店、梨膏糖商店等。

Featured Commercial Area 特色商贸区

Yuyuan Old Street

Yuyuan Old Street is located at the north of Yuyuan Tourist Mart. It is over one hundred meters long with over 30 featured shops of national customs. It gathers traditional featured products of Shanghai old Chenghuang Temple area, old famous shops and unique specialized households. Yuyuan Old Street is always bustling in the temple fairs and holidays.

豫园老街

豫园老街位于豫园商城北端，南连四海闻名的豫园，北与万商云集的福佑路小商品市场相衔，全长百余米，共有30余家极富中国民族风情的特色商铺，街内有居家用品、工艺礼品、喜庆用品和传统特色商品四大板块。豫园老街集中荟萃了上海老城隍庙地区的传统特色商品，充分反映了"名、特、优、精"的经营特色，街内既有历史悠久、渊远流长的中华老字号（如王大隆、丽云阁等），又有全国独一无二的特色专业户（如筷子店、手杖店等）。万余种传统商品汇集一市，数十家特色商店同处一街，每逢庙会假日，熙熙攘攘，摩肩接踵，呈现出一幅"吃、玩、带"的民俗画。

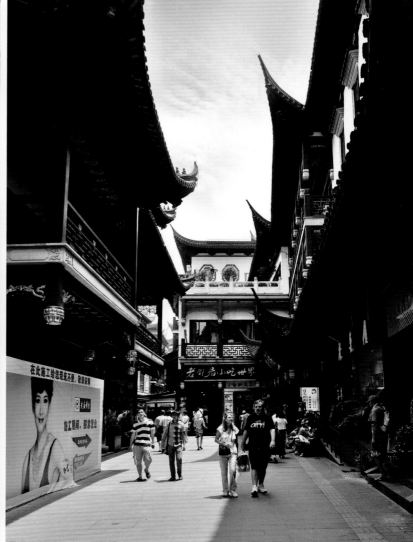

Jewelry 珠宝类

Laofengxiang

Founded in 1848, Laofengxiang is an old famous national brand of over 160-year history. By inheriting its history, improving its culture deposits, and making breakthrough in products and services, it has become one of the leading enterprises of jewelry in China.

老凤祥

　　老凤祥创立于1848年，是有着160余年历史的老字号民族品牌。老凤祥通过对其所拥有的历史和文化底蕴的传承和创新，不断地实现产品与服务上的突破，现在已发展成为中国珠宝首饰业的领军企业之一，拥有着诸多的荣誉——中国驰名商标、中国品牌产品、中华老字号、全国满意企业……

Shanghai Old Temple Gold

Shanghai Old Temple Gold is the first gold retail in Shanghai after the State Council ratified recovering the golden accessories selling in China in 1982. It originates from "Shanghai Old Chenghuang Temple Craft Store" and after decades it becomes a national well-known brand.

老庙黄金

上海老庙黄金银楼是在1982年国务院批准国内恢复销售黄金饰品后、在上海开设的第一家黄金零售点,其前身是"上海老城隍庙工艺品商店"。经过几十年的发展,老庙黄金已经发展成为一个全国知名的品牌公司。

Shanghai First Asia

Shanghai First Asia is a wholly-owned subsidiary of Shanghai Yuyuan Tourist Mart Co., Ltd. It is well-known for its gold, platinum, jewelry, jade and diamond accessories. After years of development, its line of business expands from gold ornament selling to value-added services including design, process and trade.

亚一金店

上海亚一金店是上海豫园旅游商城股份有限公司下属的全资子公司,以经营黄金、铂金、珠宝玉器、钻石首饰而闻名遐迩。经过多年的发展,它的业务范围从单纯的金饰销售拓展为集设计、加工、交易为一体的增值服务企业。上海亚一金店致力于打造"中国婚庆珠宝首饰第一品牌",获得了"中国驰名商标"、"中华老字号"、"中国黄金首饰驰名品牌"等百余项荣誉。

Food & Beverage 餐饮类

Shanghai Classical Hotel

Shanghai Classical Hotel was founded in 1875 by Zhang Huanying. For excellent cooking skills of chefs of different times and affordable prices, it has been more and more popular among citizens.

Shanghai Classical Hotel reconstructed in 1993 has a gorgeous appearance. It has three floors and is over one thousand square meters. There are all varies types of boxes such as European style and 1930s' Shanghai Gate style on the third floor. Old pictures of Shanghai in different years are hanged on the wall to remind the memories of old Shanghai.

Its dishes are also very unique. Materials must be fresh and alive; chefs must be good at all kinds of Chinese cooking skills. Their dishes are not only good looking, but also extremely delicious, therefore it enjoys tremendous popularity.

上海老饭店

上海老饭店始创于清光绪元年（1875年），原名"荣顺馆"，创始人张焕英。由于历代掌勺者烹调技术高超，且物美价廉，它很受市民欢迎，后来逐步发展壮大，成为今日的上海老饭店。

上海老饭店自1993年改建后，外观富丽堂皇，古朴典雅。饭店共有三层，营业面积1 000多平方米，三层全部为各式包房，包房设计为各种风格，有欧式包房、仿20世纪30年代石库门式包房等，在包房的命名上也各有特点，在每个楼层的墙面上悬挂着上海不同阶段的老照片，尤其是三层的包房区域，借用有轨电车和旧时招牌等道具勾起人们对老上海灯红酒绿的回忆。

老饭店的菜肴也十分有特色，烹制菜肴必用鲜活原料，在烹制手法上擅长红烧、生煸、炒、炸、煨、蒸等多种方式，注重刀工、火候，烹制成的菜肴色泽光艳、浓油赤酱、香鲜可口，深受食客欢迎。老饭店先后曾接待过印尼总统和俄罗斯总统普京夫人等名流。

Mid-lake Pavilion Teahouse

Mid-lake Pavilion is located at the middle of Jiuqu Bridge on Lotus Pond of Chenghuang Temple. It has a history of over 200 years.

In the late Qing Dynasty and early Republic of China, teahouses sprang up. Mid-lake Pavilion Teahouse together with Nimou Pavilion and Simei Pavilion composed a unique tea market scene in Chenghuang Temple.

Now the Mid-lake Pavilion Teahouse is well repaired and looks like new. It becomes a good place for tourists to have a rest and taste some tea.

湖心亭茶楼

湖心亭坐落于城隍庙荷花池上九曲桥的中心，原由明嘉靖年间四川布政司潘允端所构筑，属于豫园的内景之一。清乾隆四十九年，布业商人祝韫辉等人集资将其旧址改建成湖心亭，迄今已有200多年历史，现在的九曲桥与湖心亭旁的二层阁楼式建筑为清宣统年间商人刘慎康修建。

清末民初，茶馆业兴旺，湖心亭与城隍庙附近的凝晖阁、四美轩等茶楼构成了有城隍庙特色的茶市风光，当时上海名士王韬曾在《瀛壖杂志》里写湖心亭一带风光时说："园中名肆十余所，莲李碧螺、芬芳欲醉；夏日卓午，饮者杂沓。"更有诗人孙家振写诗赞曰："湖亭突兀宛中央，云压檐牙水绕廊。春至满阶新涨绿，秋深四壁暮烟苍。窗虚不碍兼葭补，帘卷时闻荇藻香。待到夜来先得月，俯看倒影入银塘。"

现在，经过修缮的湖心亭整旧如新，成为游客休息观景、文人雅士品茗赏趣的好去处。

Hefenglou

Hefenglou assembles the advantages of different foods in Chenghuang Temple and brings in eight big cuisines of the nation. Hefenglou has two floors. The first floor mainly provides Chinese food and the second floor mainly offers snacks of home and abroad.

和丰楼

和丰楼汇集了城隍庙饮食之大成,并引进全国八大菜系十六帮144个菜点、小吃。和丰楼分上下两层,一楼以中华美食为主,有广州风味、浙江风味、上海风味、台湾风味等300余种;二楼餐厅以海内外小吃为主。

Jiushilou

Jiushilou is located at the east side of Yuyuan small garden scenic spot. It used to be a two-floor building with the first floor as grocery and the second floor as dwelling. After transforming, now it is a DQ ice cream shop.

DQ (Dairy Queen) was founded in 1940. Now it has about 8,000 chain stores all over the world. It is an ice cream professor with the world's top sales volume.

九狮楼

九狮楼位于豫园小花园景区东侧,原是一座两层建筑,底层是杂品店、皮鞋店和铁画轩,上层是居民住宅。经过改造后,现在九狮楼是DQ冰激凌店。

DQ(Dairy Queen)意为"冰雪皇后",1940年创立,创始人麦卡洛。现在DQ已在全球数十个国家开了近8 000家连锁店,是世界销量第一的软冰激凌专家和全国连锁快餐业巨头。

Lvbolang Restaurant

Lvbolang was founded in 1929. It was expanded to become "Lvbolang Restaurant" after the first developing phase of Yuyuan Tourist Mart in 1991. After years of hard work, its products and services have earned much praise.

绿波廊酒楼

绿波廊创立于1929年，1991年豫园商城开发一期工程后，绿波廊扩大为"绿波廊酒楼"。经过多年努力经营，绿波廊酒楼形成了上海菜、上海点心、蟹宴、鱼翅四大系列，精美的餐点和温馨的服务广受赞誉。

Songyunlou

Songyunlou Restaurant is a famous time-honored enterprise. Featuring birthday party and family feast with cuisines of Jiangsu, Zhejiang, it is a comprehensive and multi-functional characteristic restaurant enterprise.

松运楼酒家

松运楼酒家是一家负有盛名的老字号，店中主要是江、浙、泸菜系，擅长特色寿宴、特色家宴，是一家全方位、多功能、具有个性特色的餐饮企业，最多可容纳四百人就餐。

Curio 古玩类

Huabaolou

Located at the west side of Chenghuang Temple, Huabaolou's building area is about ten thousand square meters. With the operation concept of "gathering good products of nature's treasures", it brings in many distinctive and unique commodities.

华宝楼

华宝楼坐落于上海城隍庙西侧，建筑面积近万平方米，由史学界泰斗周谷城先生亲笔题写的烫金楼匾光彩熠熠。华宝楼以"物华天宝汇一楼"的经营理念，在经营销售过程中引进了许多特色商品。

Department Store 百货类

Yixiulou

Yixiulou is the tallest and biggest imitated commercial building of Ming and Qing dynasties style in Shanghai at present and Yuyuan Tourist Mart is right in Yixiulou. The first floor of Yuyuan Tourist Mart is a brand fashion apparel store; the second floor is a brand lady cloth store; the third floor is a brand apparel discount store; and the underground floor is a sports and leisure wear store.

挹秀楼

豫园百货商场就开在挹秀楼中，挹秀楼是目前上海最高、最大的明清风格仿古商业建筑。豫园百货一楼为品牌时尚服饰馆，二楼为品牌淑女服饰馆，三楼为品牌服饰折扣馆，地下为运动休闲服饰馆。

Chenghuang Temple 城隍庙

Located at Chenghuang Temple Tourism Area, Chenghuang Temple is an important Taoist temple in Shanghai. It was built during Yongle Period of Ming Dynasty and now it is managed by Taoist priest of Zhengyi Faction. The temple enshrines and worships two city gods Qin Yubo and Huo Guang. The temple we can see now was the one reconstructed in 1926.

城隍庙坐落于城隍庙旅游区内，是上海重要的道教宫观。它始建于明代永乐年间，现由正一派道士管理。庙内供奉着秦裕伯和霍光两位城隍神。现在所见到的城隍庙是1926年重建而成的，殿高16米，进深21.1米，钢筋水泥结构，彩椽画栋，翠瓦朱檐。现拥有霍光殿、甲子殿、财神殿、慈航殿、城隍殿、娘娘殿、父母殿、关圣殿、文昌殿9个殿堂，总面积约2 000平方米。

Yuyuan 豫园

Yuyuan was built in 1559 and is located at the northeast of Shanghai old city area. It used to be a private garden of Pan Yunduan and is the only one garden of Ming Dynasty remained in Shanghai old city area. With elaborate design, smart layout and exquisite decoration, this building presenting the architectural art style of Ming and Qing dynasties is a treasure of Jiangnan classic garden.

豫园始建于明代嘉靖年间（1559年），位于上海老城厢东北部，原属潘允端的私人花园，是上海老城厢地区仅存的明代园林。豫园占地30余亩，园内楼阁参差，山石峥嵘，湖光潋滟，有"奇秀甲江南"之誉。现有九狮轩、得月楼、玉玲珑、积玉水廊、听涛阁、涵碧楼、古戏台等亭台楼阁以及假山、池塘等40余处古代建筑，设计精巧、布局细腻，以清幽秀丽、玲珑剔透见长，具有小中见大的特点，体现明清两代南方园林建筑艺术的风格，是江南古典园林中的一颗明珠。

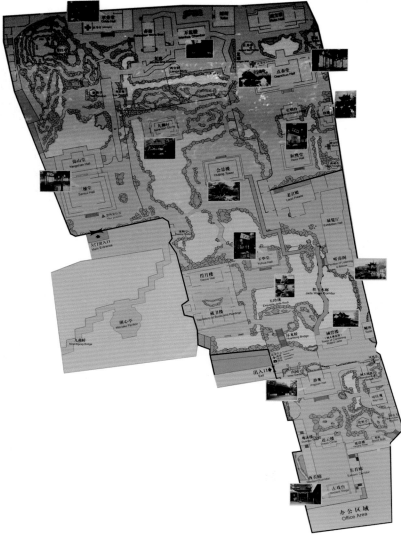

Modern Imitated Commercial Streets of Ancient Style

Yancheng Chunqiu Tourist Area, Changzhou
常州淹城传统商业街坊

街区背景与定位
Street Background & Market Positioning

History 历史承袭

Yancheng was built in the late Chunqiu Period and is of over 2,900 years history. In 1935, archaeologists in China did a field investigation in Yancheng for the first time and affirmed that it was an activities relic of ancient residents. In the 1950s and 1960s, three dugout canoes and a batch of bronze wares, potteries were unearthed which showed the cultural connotation of Yangcheng relic for the first time. After years of study, Lu Jianfang, the associate researcher of Nanjing Museum, believes that Yancheng is a military castle of Wu State which is the only well-preserved military installations of Chunqiu Period that has been found in the world.

淹城建于春秋晚期，距今已有2 900余年历史。1935年，我国考古学者首次对淹城进行了实地调查，确认淹城为一处古代居民活动遗址。20世纪五六十年代，淹城内河出土了三条独木舟和一批青铜器、陶器，这批成组文物的出土，首次展示了淹城遗址独特的文化面貌和内涵。1986年，江苏省淹城遗址考古发掘队首次对淹城遗址进行了为期6年的考古发掘，作为当年考古发掘队队员之一，南京博物院副研究员陆建芳经过多年研究后认为："淹城是吴国的一个军事城堡，是目前已发现的、世界上春秋时期唯一保存完好，有三道城河、三道城墙形制的军事设施。"

Location 区位特征

Yancheng Chunqiu Tourist Area is located at the east of Chunqiu Yancheng relic. It is composed of Traditional Chinese Medicine Street, Culture Street, and Food Street. All the three streets are facing harbor and back to river with waterside pavilions, revetments, stone bridges, corridors and so on. The whole city is surrounded by three rivers. Yancheng Chuqiu Tourist Area is a tourist scenic spot gathering modern commercial and traditional culture.

淹城传统商业街坊位于春秋淹城遗址的东部，仿汉唐式建筑，总建筑面积45 000平方米。由中医街、文化街、美食街三条古文化街组成。三条古街均临港背河，有临河水阁、水墙门、旱踏渡、长驳岸、石河桥、河埠廊坊等建筑。整座城被里外三道河流环绕。从里向外，子城、子城河、内城、内城河、外城、外城河，三城三河相套，是集现代商圈和传统文化于一体的旅游景区。

Market Positioning 市场定位

With museum, collection gallery, folk art gallery as its main body and local food, snacks, crafts, curios and local specialties as feature, the Yancheng Chunqiu Tourist Area is positioned as a comprehensive cultural commercial area assembling collection, exhibition, food and leisure.

街区是以博物馆、收藏馆、民间艺术馆为主体，以地方美食、小吃、纪念品、工艺品、古玩、地方特产为特色，集收藏、展示、餐饮、休闲于一体的综合文化商业区。

Development Concept / 开发理念

With "awaking historical memory, reproducing humanities scene, promoting traditional culture and developing leisure economy" as key concept, "a destination of cultural leisure tourism relying on Chinese Chunqiu culture" as aim, the "culture origin" function of Yancheng national relic is carried forward to the largest extent. Land utilization demands of different peripheral protection circles of the relic are smartly used to form a reasonable functional layout, a rich product system and an optimized income structure.

淹城传统商业街坊以"唤醒历史记忆，再现人文情景，弘扬传统文化，发展休闲经济"为核心思路，以"一回走千年，春秋看淹城"为形象，以建设"中国春秋文化品牌为依托的文化休闲型旅游目的地"为目标，以"内虚外实，有效保护，文化延伸，产品配套"、"烘托迷城形象，丰富文化体验，提升休闲环境"为技术路线，最大限度地弘扬淹城国家级遗址作为"文化原点"和"文化渊源区"的作用；巧妙地利用遗址外围不同保护圈层的土地利用要求，由内而外延伸出不同种类的、经过系统化的春秋文化产业内容，从而在整个街区范围内，形成合理的功能布局、饱满的产品体系和优化的收入结构。

Development Mode / 开发模式

To Present a Big Chunqiu on the Basis of Small Yancheng.
On the basis of Yancheng relic, Chunqiu culture is deeply explored. A tourism zone showing Chinese Chunqiu culture is created by presenting Chunqiu culture and to develop Yancheng tourism.

To Highlight Yancheng's Value by Bird's-eye View.
The most stunning point of Yancheng is its aerial view of its layout of tri-cities and tri-rivers. Therefore, appreciating Yancheng from the air can not only feel the stunning earth art, but also highlight the key point of value of Yancheng.

To Excite Chunqiu Culture and Promote Relic Protection.
Culture is the soul of tourism as well as the support of development of tourist destination. The development of Yancheng on the basis of exploring local culture, creates a Chunqiu culture experience park.

立足小淹城，演绎大春秋。
开发立足于淹城遗址，深刻挖掘遗址所蕴涵的春秋文化。提出"一回走千年，春秋看淹城"的目标和形象口号，以表现春秋时代文化为重心，通过淹城旅游的开发，打造展示中国春秋文化的旅游聚集区。

空中看淹城，凸显淹城价值。
淹城最大的震撼力在于——从空中鸟瞰时呈现的三城三河布局，形状酷似"眼睛"。因此，从空中观赏不仅能看到有震撼力的大地艺术——"地眼"，而且能凸显淹城价值的重点。

春秋文化活化，促进遗产保护。
文化是旅游的灵魂，也是旅游目的地持续发展的支撑。开发以淹城为依托，在深入挖掘本土文化的基础上，打造一个以春秋文化为主题的春秋文化体验园。

Street Planning 街区规划

Planning & Design Features 规划设计特色

This area is planned according to "dynamic and static divisions" and "emotional line". By arranging functional zones, an "emotional line" is formed to enable visitors to feel the extensive and profound culture of Chunqiu gradually as they walk in. For example, "the hundred schools of thought" theme zone is planned next to the entrance square to introduce the connotation of Chunqiu culture to visitors comprehensively. And then specific theme zones are planned along the visiting route.

At its entrance area, pedestrians and vehicles are divided. The vehicle entrance is the main entrance. In front of the main entrance is a landscape road and in front of the pedestrian entrance is a pedestrian road. The two roads are two axes and intersect at the evacuation square in the middle. This layout form visually and sensuously brings visitors shocks and makes visitors desire to enter the scenic spot.

The parking lot near to the main entrance is designed with some Chunqiu culture symbols, such as patterns, small imitated wares and celebrated dictum of Chunqiu Period, to create a parking lot with Chunqiu culture breath and also a "preheat" space.

设计师按照"动静分区"以及塑造"感情线"的原则对街区进行规划。通过功能区块的布置，形成了一条"感情线"，它使游客随着游览的深入，逐渐感受到春秋文化的博大精深，使其心情逐步进入高潮。如紧邻入口广场规划"诸子百家主题区"，即通过主题区各项产品的安排，向游客全面介绍春秋文化的内涵，使游客对春秋文化有个大体的认识。然后随着游览的深入，沿线规划有各个针对性的主题区。

在入口区中，行人交通和车行交通相互分离，也因此形成了车行和人行两个入口，其中车行入口是整个旅游区的主入口；正对主入口是一条景观大道，正对行人入口是一条游人步行通道。两条大道构成两条轴线，相交于中间的集散广场。这种布局形式，不仅从视觉上，也从感觉上给游客带来震撼，使其有一种迫切进入景区一睹为快的欲望。

在主入口区附近规划有停车场。规划不仅对停车场进行了生态化处理，还设计了一些春秋文化符号（如纹饰、小型仿春秋的器物、春秋名言等），从而形成一处具有春秋文化气息的停车场，同时也是"文化预热"的空间。

Street Design Features 街区设计特色

Firstly, culture, proverbs and allusions are researched and analyzed systematically, and then the significant culture, such as the hundred schools of thought, five hegemonies, bronze inscriptions and so on, is picked as objects to be created as products which are divided into landscape zone, perform zone and experience zone. Therefore, Chunqiu culture and stories are transformed into touchable tourism products.

Secondly, in the innovation of transforming Chinese ancient culture into street culture, the experience mode of combining theme park and Chinese culture creates culture theme street which combines traditional culture with modern entertainment.

Meanwhile, major culture and allusions spread so far are transformed into interaction experience project of sightseeing and entertainment. For example, Confucian culture is presented by Confucian school; story scene of Chunqiu beauties are presented by sculptures and paintings on massifs; and acousto-optic water effects are used to present theme water show.

首先，对春秋战国的文化架构、成语典故、核心文化进行系统梳理、研究、分析和提炼，挑选了诸子百家、春秋五霸、青铜铭文、春秋丽人、春秋版图等对当代影响较大的文化作为打造对象，并将产品分为春秋文化长廊板块（景观区）、春秋演艺板块（演艺区）和春秋文化体验板块（体验区），使春秋文化、春秋故事转化为可触摸、可感受、可体验的旅游产品。

其次，在中国古文化转化为街区文化的创新上，主题公园和中国文化相结合的体验模式创造了一个传统古文化与现代娱乐相融合的文化主题街区。

同时，把春秋的主体文化、最重要的文化、流传至今的典故转化为观光、游乐、娱乐、互动等体验项目，如用孔子学堂的方式来展现儒家文化、用雕刻绘画的形式在山体中展现富有传奇色彩的春秋丽人的故事场景、用春秋版图的方式表现春秋历史地理的文化内涵、用歌舞升平来包装旋转木马、用声光水影效果演艺主题水影秀等。

Major Commercial Activities 主要商业业态

Culture Street

Centralizing Yancheng Museum and Wujin Celebrities' House, cultural leisure projects such as curio, folk craft, local specialty, folk art club and teahouse, are created. There are totally 25 art and craft shops holding various calligraphy and painting exhibition, collection exhibition, master communication activities. Xiao Jianbo Art Gallery, Xu Bingyan Sculpture Research Institution and so on are emerging culture enterprises gathering art creation, appreciation communication, exhibition and collection.

Food Street

Based on Wujin folk cuisine, this street absorbs featured food from all over the country. There are 16 enterprises including restaurants offering "ten peasant family signature dishes" and "eight peasant family specialties" apart from big hotels like Sheraton and Yancheng Guild. There are Chunhui Ge, a western food restaurant which imports French cuisine and uses local materials, "Duobao Health Maintenance Restaurant" which is featuring herbal cuisine, "Changzhou Sesame Cake Shop" of over 150 years' history and so on.

Traditional Chinese Medicine Street

There are 20 various clinics, well-known drugstores, health clubs. It gathers 8 national level and 18 provincial level famous TCM doctors, more than 30 superior TCM professors with 32 folk secret recipes with exact curative effect.

文化街

文化街以淹城博物馆、武进名人馆为核心，搭配古玩、民间工艺品、土特产、曲艺会馆、茶馆等文化休闲项目，共有25家工艺美术品商店，每年举办各类书画展、藏品展、大师交流等活动20多场次。萧剑波艺术馆、徐秉言刻艺雕塑研究所、陈桂芳艺苑雅集、许家宝钟表文化艺术馆、杨纯常州工艺美术精品馆、黄洪德贞德轩、朱必清一音阁、吴京山璞玉馆、倪允晋晋伯斋、毛怀青烙画精品馆、徐国荣聚缘轩、李宏达崇雅堂、春秋画苑、戏曲会馆、淹城古玩阁、淹城美术馆……都是集艺术创作、鉴赏交流、展销收藏于一体的新兴文化企业。

美食街

美食街以武进民间美食为基础，吸纳来自全国各地的特色美食，共有经营企业16家。除了"喜来登"、"淹城会馆"等大酒店之外，美食街集聚了经营武进"十大农家招牌菜"和"八大农家特色菜"的店馆，有西太湖美食馆、芙蓉鲜螺馆、郑陆羊肉馆、戴溪青鱼馆、寨桥老鹅馆、太湖三白馆、泰村鱼圆土菜馆、湟里红星牛肉馆、焦溪二花脸菜馆。还有引进法式西餐烹技、选用本地特色原料的西餐店"春晖阁"，以家常药膳砂锅、猪羊宫宝等养生菜为特色的"多宝养生菜馆"，另有150多年历史的"常州麻糕店"等多家百年老店。

中医街

中医街设有各类特色门诊、知名药店、养生馆20家，汇聚了8位国家级名中医、18位省级名中医、30余位高级中医专家和32个疗效确切的民间秘方。有武进淹城中医门诊部、武进中医医院"国医堂"、庆和周氏中医门诊部、老人山程氏中医门诊部、中华同仁堂、兰草堂、台湾高唐中医、深圳延养堂等知名中医诊所。

Food & Beverage 餐饮类

Chunqiu Teahouse

Chunqiu Teahouse is a theme restaurant featuring tea culture. It integrates Chinese classic culture and tea culture. Its style is delicate and warm combining nobility with dignified business life smartly.

春秋茶楼

春秋茶楼是一家以茶文化为主的主题餐厅，融入了中国古典文化和茶文化，格调清馨，将高贵的气质与现代商务生活紧密结合。

Yancheng Household

Lofty building, classic environment and luxury decoration reveal a noble breath. Its featured dishes of unique taste include over-fired hen soup, Taihu Lake aquatic products, whitebait fried stick and so on.

淹城人家

淹城人家中巍峨的楼宇，古典的环境，电视中常见的王座以及座前仙鹤延年、盘龙的柱子，巨资打造的锦绣山河屏风，无一不透露着尊贵的气息。特色菜肴有老火母鸡汤、太湖一网鲜、仔乌烧萝卜粉丝、特色银鱼油条等，风味独特。

Lotus Restaurant

Lotus restaurant is a comprehensive restaurant featuring farm food. It specializes in ten peasant family signature dishes. Its operation concept is "health" and its destination positioning is "good taste, reasonable prices and dignity". Its building is in simple, graceful and antique style, and it is at an advantage ous location.

芙蓉饭庄

芙蓉饭庄是家以农家菜为特色的综合性饭店，专营湟里牛肉、雪山草鸡、芙蓉螺丝、横山桥百叶等农家招牌菜。饭店以"健康"为经营理念，以"好吃、实惠、有面子"为最终定位。建筑为简单、朴素、大方的仿古造型，店铺拥有极佳的地理位置。

The Taihu Three White Restaurant

This restaurant features "The Taihu Three White" (white fish, white shrimp and silver fish). Its interior decoration is antique and elegant with Jiangnan features.

太湖三白馆

太湖三白馆以"太湖三白"（白鱼、白虾、银鱼）为主菜，白虾因洁白透明、晶莹如玉而得名，清《太湖备考》上就有"太湖白虾甲天下，熟时色仍洁白"的记载。店内装潢古朴典雅，具有江南水乡特色。

Services 服务类

Xiayang Education

Xiayuang Education is founded by overseas returnees with abundant financial resources. It engages in consultation and transaction of overseas studying, foreign language training, and immigration.

夏洋教育

夏洋教育由海归人士创建，资金雄厚，主要从事出国留学、外语培训、移民定居等事项的咨询与办理。

Ligongdi, Suzhou
苏州李公堤

街区背景与定位
Street Background & Market Positioning

History 历史承袭

Ligongdi, which is located at Jinji Lake rim commercial district of Suzhou Industrial Park, is a long causeway in Jinji Lake.

During 1890 to 1891, a big rainstorm caused a great trouble to people in Yuanhe County. Li Chaoqiong, the magistrate of the county, organized people to build a long causeway and plant thousands of willows to reinforce it. The causeway facilitated the agriculture and traffic between Yuanhe County and east side of Jinji Lake, therefore, the area around the causeway became prosperous. In order to memorize Li Chaoqiong, the causeway is named after him, called Ligongdi.

The street is generally delicate. Visitors can know about its spectacular scene in old days from the ancient inscription on the causeway. Landscape of the street is designed by EDAW Earthasia, Ltd., one of the leading landscape designers and regional planning consultants in the world. Now the street is a successful example of comprehensive aquatic community in the country.

李公堤位于苏州工业园区的环金鸡湖商圈，是金鸡湖的湖中长堤。

公元1890—1891年年间，正遇天灾，阴雨成涝，时任苏州府元和知县的李超琼决定以工赈灾。他组织百姓将太平天国运动期间被毁的民房等建筑垃圾运到金鸡湖中，筑成长堤，并在堤上种植杨柳数千株以加固堤防。自此长堤稳固，城东民田滋润，湖水潋滟，并且出现了商贾云集、盛世繁华的景象。百姓为感谢李超琼，便把此堤称为"李公堤"，时人还将它比做杭州西湖的白堤和苏堤。李公堤的建成，大大便利了当时元和县与金鸡湖东部地区的交通，斜塘也因此逐渐繁荣，乃至形成集镇。

街区整体典雅精致，当年商贾云集的盛景，从堤上古碑文中仍可略窥一二。现李公堤景观由全球领先的景观设计与区域规划顾问之一——美国泛亚易道公司设计。其现已成为国内混合型亲水社区的成功典范，形成了规模巨大的开放型现代城市生态公园。

Location 区位特征

Ligongdi is located at the core area of Suzhou CBD which is under construction. There is a central financial business center and several star hotels around it. There are also many mature commercial projects such as Left-bank Commercial Street, Lakeside Xintiandi, Era International Shopping Mall, and a international golf course with most completed equipments in Eastern China and other dozens of upscale clubs. And its traffic is rather convenient with several bus routes gathering here.

李公堤所在商圈是建设中的苏州CBD核心区域。周边有中央金融商务中心、星级酒店如新苏国际五星级酒店、金鸡湖大酒店、尼盛万丽国际酒店、中茵皇冠度假酒店等；诸多成熟的商业项目如左岸商业街、湖滨新天地、苏州首座"Shopping Mall"——时代汇国际街区、湖滨U形广场等；还有华东地区设施最完备的国际高尔夫球场及数十个大规模的高尚社区。多条市区公交线路汇集，沪宁高速公路、苏嘉杭高速公路及机场可快速便捷通达。

Market Positioning 市场定位

Ligongdi is positioned as an international local style commercial aquatic street gathering upscale specialty catering, entertainment, tourism and culture. It assembles various famous brands from Denmark, Italy, Germany, Japan, Belgium and Chinese Hong Kong.

李公堤定位为集高端特色餐饮、娱乐、观光、休闲、文化为一体的国际性风情商业水街，汇聚了来自丹麦、意大利、德国、日本、比利时及中国香港等地的知名品牌商家，于2006年12月开街。

规划设计特色
Planning & Design Features

Street Planning 街区规划

The planning of the street has paid great attention to enhance cultural connotation and enrich cultural activities. Therefore, the developing concept of combining business, tourism and culture is realized. Emphases are laid on the culture of Suzhou, meanwhile relative culture and features of home and abroad are shown to make Ligongdi international.

A comprehensive street of multiple commercial activities including business, culture, idea, entertainment, exhibition and shopping is completed. Opened sightseeing, leisure holiday, healthy life and business communication are integrated here. This is a top waterfront leisure commercial street and a new economic area of fashion, business and culture.

项目总占地面积324 600平方米，总建筑面积约258 000平方米。规划注重提升文化内涵，丰富文化业态，使商、旅、文结合的开发理念成为现实。在引进传统文化的同时，项目注重多元文化的丰富性，实现动与静的结合，传统与现代的结合，打造风尚、风情、风味的现代生活新模式。在注重苏州文化的同时，设计师也注重展现国内外相关文化、特色，让李公堤不仅是苏州的李公堤，也是中国的李公堤。

建成后，此地形成了涵盖商务、文化、创意、娱乐、休闲、展览、购物等的都市多元业态组合街区，成为开放式游览、休闲度假、健康生活、商务交际等相互交融之地，时尚与繁华并重的苏城经济新领地，商业与文化并举的苏州首席滨水景观休闲商业街区。

Street Design Features 街区设计特色

Ligongdi organically combines the Jinji Lake with modern multi-style, history with reality, tourism with business. Now it has become one of the most popular areas of the most successful operation in Suzhou.

The design paid attention to both excavation and import means to explore the local culture and create cultural landscapes, and to enrich the culture of the street by importing other characteristic cultures.

First phase — the street is developed on the base of history. Several bridges linking up islands, layout of antique garden of Suzhou, and grey, white and black as three main color tones show the charm of Jiangnan. The style of buildings keeps the Suzhou traditional residential form. International well-known brands and old Chinese restaurants are gathered here.

Second phase — European style buildings smartly integrates with neo-classic paths, bridges and pavilions to create a unique modern Suzhou waterfront neighbors. Various kinds of well-known bars add new experience to the night life of this area.

Third phase — a European style commercial street connects all business activities to form a multi-functional featured commercial street group. The concept of "street mall" is firstly brought to Suzhou. The location advantage and featured buildings are fully used to build a comprehensive service community gathering leisure, entertainment and shopping.

Fourth phase — it is the last area to be developed in eight sights of Jinji Lake rim. It is a unique featured peninsula lake shore commercial area which is a brand new commercial district assembling retail, restaurant, leisure and entertainment.

李公堤通过"桥堤文化"和"湖滨公园"把金鸡湖的水、绿与姑苏的文化结合在一起，将金鸡湖与现代多元风情、历史与现实、休闲旅游与商业有机地组合起来，已成为苏州地区最有人气、商气且运营最成功的区域之一。

设计以挖掘与引入并重，不仅对李公堤历史人文进行了深入挖掘，完成了李公堤碑亭、李超琼书画、李超琼雕塑及诗碑、《李公堤记》碑文等文化景观的营造，还引入多元个性文化，为街区文化不断"输入灵魂"。

一期：在历史的基础上开发，多座小桥串连各岛，苏州古园林式布局，灰、白、黑三色为主色调，尽显江南水乡神韵。建筑风格保留了传统的苏州民居形态，汇聚了包括国际的著名品牌以及中华老字号在内的知名餐饮商家。

二期：欧陆风尚建筑与新古典主义道桥亭台巧妙结合，营造出苏州现代天堂之水巷邻里的氛围。与"南京1912"合作打造李公堤"1912酒吧街区"，集中了国际国内知名慢摇吧、迪吧、纯饮吧等各类酒吧，为园区的夜生活带来了新的体验，成为苏州首席休闲酒吧娱乐动感区。

三期：一条欧陆风情的商业步行街将项目所有的业态串连起来，形成一个多功能的特色商业街组合，将"Street Mall"的概念首次带入苏州，利用优越地段以及特色的风情建筑，打造集休闲、娱乐、购物为一体的综合服务街区。

四期：环金鸡湖八大景观中最后一块待开发区域，是集零售、餐饮、休闲、娱乐等为一体的半岛临湖绝版特色商业领地，旨在全力打造金鸡湖新商圈。

Modern Imitated Commercial Streets of Ancient Style

Major Commercial Activities 主要商业业态

Ligongdi International Aquatic Street is divided into dynamic part and static part. The dynamic part includes mainly theme bars, music restaurants, discos and featured restaurants. This part drives the popularity all day long and is the most active area at the south bank of Jinji Lake. The static part targeting at business group and VIP assembles upscale restaurants, spas, business clubs and leisure hotels.

Ligongdi perfectly shows a modern commercial street with profound history and culture, honesty business and operation, and well-known brands. It includes international celebrated brands, as well as old famous Chinese brands. All the modern business activities insert urban expressions into Jiangnan gardens.

The commercial plan fully explores the landscape and history resources provided by Ligongdi and organically combines tourism with business to enable the street not only to provide business carrier for the tourist economy of this area, but also to make the business projects themselves well-known scenic spots. Therefore, tourist projects and business carrier are perfectly combined.

李公堤国际风情商业水街分为动、静两个区域。前者以主题性酒吧、音乐餐厅、咖啡吧、迪斯科舞厅及特色餐饮为主，该区域全天候带动人气，成为苏州金鸡湖南岸极具活力的区域。而后者则集高档餐饮、Spa生活馆、商务会所、休闲酒店于一体，主要针对商务人群和VIP人士。

李公堤将"历史有根、文化有脉、商业有魂、经营有道、品牌有名"的现代特色商业街的特征完美呈现出来。一侧是中华老字号，吴地人家、得月楼、老东吴等；一侧是世界知名品牌，北欧风情的酒吧、音乐餐厅、老农舍、意大利创意餐厅、番茄主义、慕尼黑HB皇家啤酒馆等，还有时尚水疗度假的香港御庭Spa精品酒店。无论是喧闹的主题酒吧、音乐餐厅，还是优雅的高档餐饮、Spa馆、商务会所，都能于园林水乡间找寻到都市表情。

商业策划充分挖掘李公堤所拥有的景观、历史资源，将休闲旅游与商业购物等有机组合，使街区不但为区域的旅游经济提供了商业载体，而且使商业项目本身也成了知名的旅游景点，实现了旅游项目和商业载体的完美结合。

Food & Beverage 餐饮类

Fragrant Camphor Garden

Themed creative Cantonese cuisine and Chinese food in Western style, it takes most advantage form nature, history and culture to make its building integrate with all kinds of elements, to create a special concept of "implying design in nature". It is a brand new and fashionable place for leisure.

香樟花园时尚厨房

该店以创意粤式料理、中餐西吃为餐饮主题，充分利用自然、历史、人文等丰富资源，让建筑与各种因素相互融合，创造一种"寓设计于自然"的特殊概念，为全新、时尚、休闲的好去处。

Charles's Premium Steak House & Cigar Bar

Originating from the first brand steak restaurant introduced to China from America, this restaurant uses the latest baking oven which can reach the high temperature that keep the original taste and flavor of the fresh steaks.

查理仕牛排·雪茄馆

该店源自美国第一家引入中国的品牌牛排馆，是美国高佛酒店管理公司旗下的著名餐饮品牌，店内使用美国最新的牛排烘烤炉，炉内的温度瞬间可达到千度，使新鲜的牛排保持原汁原味。

Asahi House

Inheriting the Japanese traditional features, the restaurant provides customers with authentic Japanese flavor. It satisfies different groups of customers with diversified order mode like buffet, set meals and single dishes. Its refined and relaxing environment and high-quality service add more praise to it.

朝日屋

朝日屋秉承了日本的传统特色，质朴的竹帘，明幽的日本灯，闲适的榻榻米、精致的料理，无论是感官还是味蕾，都能让人享受和风清扬的地道日本风情。该店以自助、套餐、单点等多样化的点餐模式满足多元化的消费群体，清雅悠闲的店堂环境和高品质的服务质量更为朝日屋的优良口碑锦上添花。

Dain Ti Hill

The furnishings of Dain Ti Hill are sedate and present a new fashion cultural style. Continuing the classic charm of Tang and adding modern innovative fashion elements, this restaurant, which is relaxing, interesting and modern, creates an atmosphere that perfectly integrates fashion and classicism.

代官山

店内陈设蕴涵着深思熟虑、内敛沉着的趣味，体现出新时尚人文风格。延续唐风的古典韵味，加入时代创新的流行元素，轻松、有趣、富现代感，营造出时尚流行与古典简约完美融合的氛围。

Shanghai Xiao Nan Guo

It is a national famous Chinese food chain brand of middle and top class. Representing Shanghai style cooking, this restaurant is popular among consumers in Suzhou.

上海小南园

上海小南园首次引入苏州，是国内知名的中高档中餐连锁品牌，店中作为海派美食代表的精品菜肴，深受苏州消费者的欢迎。

South Beauty

South Beauty was founded by Zhang Lan and is distributed in Beijing, Shanghai, Tianjin, Chengdu, Shenzhen, Suzhou and so on. Now it has become one of the most promising international food service management companies in China and is leading Chinese cuisine toward international market.

俏江南

俏江南品牌餐厅由张兰创立,在北京、上海、天津、成都、深圳、苏州、青岛、沈阳、南京、西安、无锡、宁波等地都有分店,如今已经成为中国最具发展潜力的国际餐饮服务管理公司之一,引领着中华美食文化走向国际市场。

Liuhuiguan

It selects top-class Yangcheng Lake hairy crabs to cook with vinegar of secret prescription. Apart from tasty and refreshing crabs, steamed chicken, poached sliced beef and boiled fish with picked cabbage and chili are also its signature Sichuan dishes.

六会馆

六会馆选用专门出口日本、新加坡等地的上等阳澄湖大闸蟹,佐以秘方熬制的醋,味道浓郁爽口,回味清甜。除了令人食指大动的蟹宴,口水鸡、水煮牛肉、酸菜鱼都是招牌川菜。另有特色佳肴如宫保腰果虾球、碧绿鱼香带子等。

Entertainment & Leisure 娱乐、休闲类

Ligongdi Yihao

Ligongdi Yihao is an international business KTV club located at the left bank of Jinji Lake. It is a tasteful, fashionable and pure business KTV designed by senior designer.

李公堤一号

李公堤一号国际商务KTV会所依偎在美丽的金鸡湖左岸，是由资深设计师精心打造的高品位、新潮流、纯商务的KTV娱乐之都。

Suisse Place

Suisse Place is a hotel brand originating from world well-known Swiss serviced apartment brand. It subordinates to Swiss-Belhotel International which is over 20 years old and has 58 hotels and projects around the world covering 13 Asia-Pacific countries and regions including China, New Zealand, Indonesia, Malaysia, Vietnam, Australia and the Middle East area.

瑞贝庭

瑞贝庭酒店源于全球知名的瑞士服务式公寓品牌,其所属的瑞雅酒店集团至今已有20多年的历史,在全球拥有58个酒店及项目,分支机构遍布中国、新西兰、印尼、马来西亚、越南、澳大利亚及中东等地区十三个国家和地区。

Shuibawang

Shuibawang features Korean style sweat steam. With a operation concept of "quality is life and soul", it provides high-quality products and after-sale service, so it develops rapidly.

水霸王

水霸王韩式汗蒸,秉承以"品质为灵魂,质量为生命"的经营理念,以过硬的产品及优质的售后服务,使企业始终保持快速增长。从生产单一的产品到泳池、桑拿、水疗的整体设计安装,实现了质的飞跃。

李公堤荣誉 / Honors

2009 — The Most ValuableReal Estate Project in Changjiang River Delta
2009 — The first "China Featured Commercial Street" in Suzhou
2010 — National Commercial Street Advance Group
2010 — "Top Ten Most Beautiful Night Scenic Spots in Suzhou"
2011 — "The Real Estate Project of Most Innovative Value in China"
2011 — "The Most Beautiful City Card" by Xinhua Net

2009年，长三角最具品牌价值地产项目
2009年，苏州市区首个国字号"中国特色商业街"
2010年，全国商业街先进集体
2010年，"苏州十大最美夜景地"
2011年，"中国最具创新价值商业地产项目"
2011年，代表苏州入选新华网，成为"中国最美的一张城市名片"

引导指示系统 / Guidance & Sign System

Ancient Culture Street, Tianjin
天津古文化街

街区背景与定位
Street Background & Market Positioning

History 历史承袭

Radiating from Tianhougong Temple of Yuan Dynasty, Ancient Culture Street is composed by imitative Qing Dynasty small shops, and is 687 meters long and 5 meters wide.

During Yuan Dynasty, great quantities of food are transported to the capital by the way of Tianjin every year. The government honored Tianhou as Goddess to pray for safeness for shipping and built Tianhougong Temples along the coastal towns, thus the Tianhougong Temple of Ancient Street was built. It was said that Tianhou was born at 23th of the third month in lunar calendar, therefore people celebrated this day. And in Qing Dynasty the celebration was named "Emperor Party", for Qianlong Emperor had once attended this celebration. After Tianjin Ancient Culture Street was opened in 1986, there are shows such as dragon lantern dance, lion dance, Beijing opera and etc. in Huanghui every year.

天津古文化街以元代古迹天后宫（建于1326年）为中心，由仿中国清代民间的小式店铺组成街道，长687米，宽5米。古街南北街口各有牌坊一座，上书"津门故里"和"沽上艺苑"。

元代期间，京城每年需从南方运入大量粮食，先海运到天津，再从天津河运至京城。当时政府为求航运平安，便尊护航女神（天后）为天妃，并在沿海城镇建起天后宫，古文化街天后宫便于此时建造。相传天后诞辰在农历三月二十三日，因而每年此时人们都会为天后举行盛大的娘娘会，后在清代时期，乾隆皇帝下江南时曾游玩此会，因而得名"皇会"。1986年天津古文化街建成开业后，每到皇会，人们都会表演龙灯舞、狮子舞、少林会、高跷、旱船、京戏、梆子、法鼓等节目。

Location 区位特征

Tianjin located at the joint of five branches of Haihe River in North China Plain has a long history and profound culture. Because of its advantageous location, Tianjin has been advanced in water transport since ancient times. After the foundation of PRC, Tianjin has become one of direct-controlled municipalities and its economy mushroomed.

The Ancient Culture Street is located in northeast corner, Nankai District, Tianjin. As one of "ten scenic spots in Tianjin", it has a long and profound history and culture.

天津历史悠久，文化底蕴深厚，位于华北平原海河五大支流汇流处，东临渤海，北依燕山，天津的母亲河海河在城中蜿蜒而过。凭借着良好的河海水脉，天津自古以来漕运发达，是京杭水运的重要交通枢纽。新中国成立后，天津成为中国直辖市之一，是环渤海地区的重要城市，经济飞速发展，呈现一片繁华景象。

古文化街位于天津南开区东北隅东门外，西接海河，北起老铁桥大街（宫北大街），南至水阁大街（宫南大街）。作为"津门十景"之一，古文化街历史文化底蕴深厚。

Market Positioning 市场定位

Ancient Culture Street is positioned as a comprehensive tourism and business zone by fully using current tourism and commercial resources to make it the most featured leisure zone for tourism and business with Tianjin folklore characteristic and cultural connotation.

天津古文化街的定位为充分利用现有旅游资源和商业资源，扩展现有的商业文化、民俗文化特色，建成集旅游、购物、餐饮、休闲、住宿为一体的综合性旅游商贸区，使之成为天津海河沿线最具天津民俗特色、最富津味文化内涵的商旅长廊和天津人感怀掌故、外埠游客了解津貌不可错过的商旅休闲区。

规划设计特色

Planning & Design Features

Street Planning 街区规划

1. General Layout
Taking Gongbei Street and Gongnan Street as axis, Ancient Culture Street extends from south to north with a memorial archway at each street entrance. Gongbei Street and Gongnan Street together with Dashizi Hutong and Wazi Hutong divide the street into commercial plaza, Tongqingli Building Group, Yuhuang Pavilion, Curio Town, Culture Mini-town, Tianhougong Temple, Folk-custom Cultural Center, amorous bank and waterside terrace and organize the major landscape of this street.

2. Business Functional Sections
Based on the principle of leading by culture, the business section is divided into three parts:
1) Ancient Culture Street with archaic charm. Mainly composed by historical sites and featured buildings, this street is full of Tianjin local culture features. There is not only Tianhou Temple, Yuhuang Pavilion the Taoist temple with 600 years old and other featured old buildings, but also newly-built about Culture Mini-town and other scenic spots.
2) Food & beverage entertainment area on waterside terrace. With waterscape and large waterside terrace, Haihe River amorous area is planned along the river which includes KTV, restaurant, entertainment center, business club, Food Street and etc.
3) Modern business center. There are mainly specialized shopping malls and super markets. And the Official Banking Firm, an old style bank operated by government, is recovered.

1. 街区整体规划
天津古文化街以宫北大街和宫南大街为轴线南北向延伸，保持了古街原有的肌理，古文化街南北街口各有一座牌坊，分别书以"津门故里"和"沽上艺苑"。宫北、宫南大街结合大狮子胡同和袜子胡同，由北至南将街区分为商业广场、通庆里建筑群和玉皇阁、古玩城、文化小城、天后宫和民俗文化馆、风情水畔和亲水平台等几个区域，组织起街区的主要景观。

2. 商贸功能区块
项目在进行土地开发总平面规划时，坚持以"文化"引领营销为原则，在挖掘传统文化、民俗文化与旅游文化、商贸文化结合的基础上，以"旅游带商业，商业促旅游"，以发展多元化经济为导向，将商贸划分为三大街区。

1) 古韵悠然的古文化街区。以街区的文化古迹和特色建筑为主，极具天津地域文化特色。区内以天津妈祖文化代表的天后宫为根基，结合经过整修的具有600多年历史的道观玉皇阁、古老民居通庆里和特色刘家大院，汇集新建的文化小城、古玩城、民俗文化馆等，几个项目交相辉映，构成商业街区的核心业态——文化街区。

2) 亲水平台的文化沿河餐饮娱乐。以亲水景观和大型亲水平台为主，沿海河规划海河风情带。区内设有量贩式KTV、餐饮、娱乐中心、商务会馆、美食街等项目。

3) 现代商贸中心。以错位经营的专业商城和专业超市为主，恢复"官银号"银行，兴建专营美容美发用品的美博城及专业超市、精品店铺等。

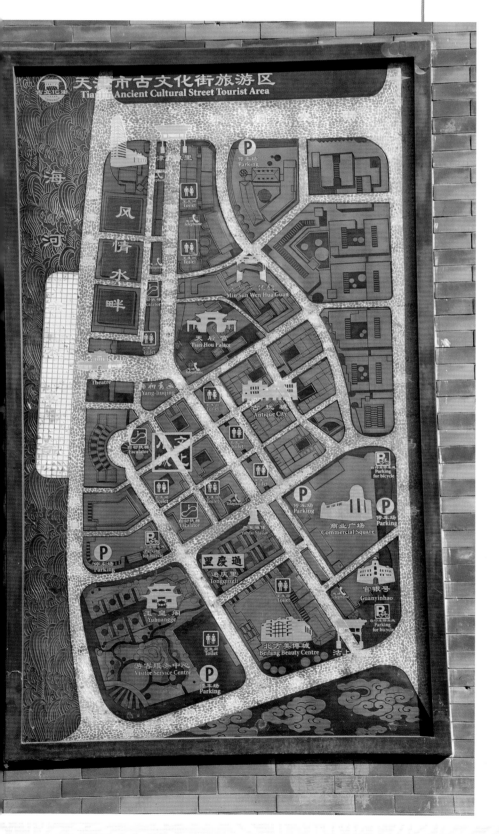

Street Design Features 街区设计特色

Ancient Culture Street covers 22,000 m² with well-arranged antique buildings. All the buildings are decorated with various partition boards, doors, windows, rails, tablets, lanterns, wood sculptures and colorful paintings. The street is paved with 12 Chinese animals' sculpture and green features, to make the commercial area take on an ancient taste.

There are lots of time-honored shops, with special food all over China and leisure places. And the Tianhou Temple, Royal Palace, and so on provide the street with modern taste.

古文化街建筑面积2.2万平方米，古建筑错落有致、蜿蜒曲折。所有的名堂，全部青墙红柱、磨砖对缝，修饰以形式多样的隔扇门窗、栏杆、屋顶翼角，古朴、典雅却不失隽秀，匾额、楹联、宫灯、旗幡、精细的木雕及上千幅色彩艳丽的油漆绘画的点缀，使街道古典文化气息弥漫。街路以"十二铜钱""十二生肖印章"铺装，在街心小品、绿色点缀、广场休憩等景观上，配合灯光效果设计，使得商贸区呈现古韵新风。

古文化街中店肆林立，既有古玩城、文化小城等古玩、古物市场，又有泥人张、刻砖刘、老美华、联升斋等民间知名商铺；既有汇集各地美食的风情美食街，又有提供休闲娱乐的亲水平台、戏楼；既有历史悠久的建筑天后宫、玉皇阁、通庆里，又有充满现代时尚气息的商业广场、北方美博城。

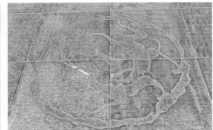

Modern Imitated Commercial Streets of Ancient Style
现代仿建的商业街区

Major Commercial Activities 主要商业业态

As one of "ten scenic spots in Tianjin", Ancient Culture Street insists on an operating way of "Chinese style, Tianjin Style, Cultural style and Archaic style". It mainly engages in cultural good. In its subsequent developing, featured commercial areas are added to the street to make it a comprehensive tourism and business zone. Its commercial activities are as follows:

1) Shops selling cultural relics, curios and antique furniture and so on, namely, Longchangxing, Daqingyoubi, Li Silversmith.
2) Folk traditional commodities such as colored clay figurines, Spring Festival pictures, kites of Kite Wei, porcelain from Jingdezhen, embroidery of Suzhou and so on. The shops are Clay Figurine Zhang, Yangliuqing Spring Festival Picture Society, Kite Wei, Brick Sculpture Liu and Xiuzhuzhai.
3) Modern commodities such as KTV, restaurant, entertainment center, business club and Food Street.

作为"津门十景"，天津古文化街一直坚持"中国味、天津味、文化味、古味"的经营特色，以经营文化用品为主。在后续的开发中，古文化街新建了民俗文化馆、文化小城、古玩城、北方美博城、北方赌石城、华夏鞋文化博物馆等特色场馆，成为了集民俗文化、旅游购物、餐饮住宿、娱乐健身于一体的综合性旅游商贸区。其商业业态的具体情况如下。

1）经营文物、古玩及古家具等商品，如文房四宝、古旧书籍、传统年画、珠宝玉石、古式家什等。具体商铺有隆昌兴、大清邮币、李银匠等。

2）民间传统商品有娃娃乐、泥人彩塑、年画、风筝魏的风筝、砖刻、修竹斋的刘海空竹、景德镇的瓷器、苏州的刺绣等，具体商铺有泥人张、杨柳青年画社、风筝魏、刻砖刘、修竹斋等。

3）现代商业包括量贩式KTV、餐饮、娱乐中心、商务会馆、美食街等。

Featured Commercial Area 特色商贸区

North Stone Gamble City

The North Stone Gamble City has been located at the east of Culture Mini-town since 2011. Its area is over 4,000 m² with over twenty stone gamble shops. The thirty eight columns and façades of shops, eaves, corridors in the North Stone Gamble City lively introduce the knowledge of stone gamble.

北方赌石城

北方赌石城占地4 000多平方米，营业面积3 000多平方米，聚集将近20家赌石商铺。赌石城内店铺门面、外檐、连廊等38个立柱面图文并茂地介绍了赌石文化等相关方面的知识。

Food Street

The Food Street is adjacent to Gongqian Square at the north. The carved façades and eaves of the shops are full of archaic charm. There are thirty five shops and fourteen snack pavilions for tourists to enjoy delicious food when they are travelling.

美食街

美食街北起于宫前广场，沿街店铺门面外檐雕梁画栋，古风十足。全街共有35家店铺，经营总面积5 000平方米左右，街面设有14个风味小吃亭。

Tongqingli

Tongqingli is a Chinese-Western style residential house group built in late Qing Dynasty and early Republic of China. It located at the northeast corner of Ancient Culture Street is a historical building group under key protection. It is a Chinese style lane composed by ten independent courtyards. There are two two-floor buildings with black bricks in each courtyard. This house group is integrated with traditional Chinese lanes layout and individual building structure of Western architecture.

通庆里

通庆里是始建于清末民初的中西合璧风格的民用住宅建筑群，位于古文化街东北隅，是重点保护的历史风貌建筑。通庆里为十个独立的院落串联而成的中式里巷，每座里巷的出入口都有过街楼，楼口上端一般会镶嵌蝴蝶状的镂空木雕，寓意"通达吉庆"。院落内建有两座二层砖木结构的青砖楼房，屋顶成坡状，楼房一、二层均有开敞式的外廊。整个建筑群融入了中国传统的里巷布局和西洋建筑中的单体建筑结构。

Culture Mini-town

Culture Mini-town is composed of five relatively independent spaces which are named after Chinese traditional "Wuxing" and apply the principle of "Wuxing" in their design.

文化小城

文化小城由五个相对独立的空间组成，以中国传统的"五行"命名五个院落，在设计中同样运用五行"相生""相胜"的原理。

Curio Town

Curio Town is composed of six relatively independent individual buildings of over ten thousand square meters. It is the biggest curio market in North China.

The Curio Town combines Chinese traditional black bricks with Western unique castle style buildings. Exquisite landscapes are designed in the six individual yards, a symbol building is built on the main street and a small square is designed on the joint of two planning roads, therefore, a garden style shopping environment is created.

古玩城

古玩城由6个相对独立的单体建筑构成，占地面积1万多平方米，是华北地区做大的古玩市场。古玩城地下一层为大型地下停车场，地上五层为经营店铺及仓储用房。主要经营古玩物品，还汇集有玉器、字画、现代文化商品、工艺品等经营项目。

古玩城将中国的传统的青砖与西方独特的城堡式建筑相结合，凝聚了中西文化的精髓。6个单独的小院中分别设计别致的景观，在主街道上建设标志性建筑，在两条规划路的交会处设计小广场，打造出花园式的购物环境。

Crafts 工艺类

Clay Figurine Zhang

Clay Figurine Zhang was founded in the Qing Dynasty by Zhang Mingshan. His works of art cover a wide area and adopt materials from myth, dramas, novels and real life. The works represent lively images and depict personalities. From generation to generation, Clay Figure Zhang has reached the peak of clay art and gained widespread support in China.

泥人张

天津泥人张创于清道光年间，创始人张明山。泥人张的作品取材广泛，多取材于神话、戏剧、小说及现实生活，通过"塑造"与"绘色"表现出生动的人物形象，"随类赋彩"地刻画出人物的性格，使作品极具生命力。经过几代传承，泥人张已成为我国泥塑艺术的代表，深得民众喜爱。

相传泥人张创始人张明山在泥塑时，只需和人对面坐谈，抟土于手，不动声色，瞬息而成。面目径寸，不仅形神毕肖，且栩栩如生，须眉欲动。

Liansheng House

Set up in 1983 by Li Jifang, the Liansheng House is famous for exquisite embroidery all over the world. Shortly after its founding, the house became attached to the Suzhou embroidery and inherited the Suzhou techniques. In 1985 when the Ancient Culture Street was renovated, the Liansheng House became well-known again. In 1989, Mr. Wu Qingcheng purchased the house and integrated the celebrity paintings into the embroidery for the first time ever.

联升斋

1983年李霁舫创立联升斋，以其精湛绝伦的刺绣工艺闻名于世。中国四大名绣以苏绣为首，联升斋创办之初便与苏绣"结缘"，传承了苏绣的精湛技艺，刺绣方面以制作彩轿、新娘服饰、百子图等为主。1985年，古文化街第一次修建，联升斋便恢复老字号，坐落于古文化街中。1989年，吴庆城先生买下联升斋，此后改变了联升斋传统的刺绣形式，将名人字画引入刺绣，使刺绣成为值得收藏品鉴的艺术佳品。

Lao Meihua

Founded in 1911, the Lao Meihua has a long history. In 1993, it was named the time-honored shop in Tianjin. Six traditional manual skills of Lao Meihua have been selected into the protection list of Tianjin and National Intangible Cultural Heritage. For almost a hundred years, Lao Meihua has gain trust and praise from consumers for its products' quality and careful service.

老美华

天津老美华创立于1911年,是享誉津门的百年老店,1993年被国家商务部命名为"中华老字号",并被国家工商总局认定为中国驰名商标。老美华的六项传统手工技艺已入选天津市和国家非物质文化遗产保护名录。百年来,老美华以其选材精良、工艺精湛、服务精细,赢得广大消费者的信赖和赞誉。

Silversmith Li

Silversmith Li was founded in 1998. It is the domestic early pathfinder of silver accessory processing enterprises of independent design. It inherits and promotes Chinese traditional culture and devotes to unique accessories for different demands of different groups.

李银匠

李银匠创立于1998年，是国内最早以自主设计为核心的个性纯银饰品加工企业的开创者。李银匠以"传播银饰文化，缔造美丽人生"为使命，以"银"为媒介，传承与发扬中国传统文化，以"匠人"自许，致力于打造适合不同人群需求的个性化饰品。

Yangliuqing Painting Shop

Yangliuqing Painting Shop in Ancient Culture Street is one of Yangliuqing Spring Festival Picture Society's points of sales. It is adjacent to Culture Mini-town in the north and Food Street in the south, next to the Tianhou Temple in the west and the Playhouse in the east. It is located at the center bustling area of Ancient Culture Street.

杨柳青画店

古文化街杨柳青画店是杨柳青画社的销售点之一，它北接文化小城，南连风味美食街，西邻天后宫，东依戏楼，处于古文化街中央繁华地带。郭沫若题名的招牌"杨柳青画店"显示出它的文化气息与历史沧桑。

Collections 收藏类

Renjunxuan

Renjunxuan which is famous throughout Tianjin and even China was founded by Song Guangren, the celebrity in field of red porcelain of Tianjin. Renjunxuan is always filled with tea aroma and is an important salon for nationwide red porcelain collectors to communicate appreciation and experience.

仁君轩

仁君轩在天津乃至全国可算是大名鼎鼎了，由天津紫砂界名人"大强"宋广仁所创。仁君轩中整天紫气满堂，茶香四溢，是天津及全国各地紫砂收藏爱好者交流鉴藏心得的一个重要沙龙。

Daqingyoubi

Daqingyoubi is located at the northeast side of Tianhougong Temple, facing Clay Figurine Zhang and Silversmith Li. Both exterior decoration and interior furnishings of Daqingyoubi are of antique beauty which reveals the classical allusion of the shop.

大清邮币

　　大清邮币位于天后宫东北侧，与泥人张和李银匠店铺相对。大清邮币黑底白字的旗幡异常显眼，很远就能看见。古朴庄重的匾额显得古味十足，进入店门，木质的桌、柜上摆放着透出历史气息的文物，配上暗黄色的灯光，古色古香。天花上遒劲的书法作品，透露出店铺的古典韵味。

Food & Beverage 餐饮类

Mei Mansion Family Feast

Mei Mansion Family Feast is located at No.5 Tongqingli, Ancient Culture Street, which is a rather secluded quadrangle courtyard. Its dish styles are all the hundreds of dishes handed down from the older generations of Mei Lanfang's family and are exquisite and extraordinarily delicious.

梅府家宴

　　梅府家宴位于古文化街通庆里一个非常隐蔽的四合院中，巷院深深，灯笼高挂，进院正面摆的是梅兰芳"贵妃醉酒"的画像，画像下面有傲雪的梅花，梅花旁边栽种着高傲不屈的竹子。走进屋内，厚厚的棉门帘隔绝了尘世的喧嚣，房内放着小曲，十分幽静。梅府家宴中的菜式全部来自梅兰芳家传数百道菜，菜品十分精致，美味异常。

Haiya Teahouse

Haiya Teahouse is situated at No.3 Tongqingli with about one thousand square meters. The stone lions and red lantern in front of the gate subtly integrate the teahouse with the Ancient Culture Street.

The main building of the teahouse is a two-floor courtyard of late Qing Dynasty, which is graceful yet lively. The tables and chairs made of rosewood with simple and exquisite carvings establish a interior main color tone, and golden wallpaper and cushions express a gentle ambience, which add visual aesthetic.

海雅茶园

海雅茶园会馆位于天津市古文化街通庆里三号院，整体面积1 000平方米左右。门前的石狮与红灯，一挑金边红底的茶幡，将茶园与古文化街的古典与质朴巧妙地融为一体。

茶园主体建筑为一座四面合围的晚清二层套院，红木为门，绿栏为廊，在青砖的淡雅之中平添了几分活泼与灵动。茶室内的红木桌椅，奠定了室内的主色调，桌椅雕花简洁而不繁华。墙壁饰以金色壁纸，将茶室内温婉柔和的氛围淋漓尽致地表现出来，与金色坐垫相呼应，在舒适的同时，增添了视觉上的美感。

Finance 金融类

Official Banking Firm

Located in the heartland of Tianjin, the official banking firm was once the official financial institution. Since 1900 when the Eight-power Allied Forces intruded Tianjin, lots of banks were set up there, causing financial chaos and the decrease in domestic financial firms. The ruler then in Tianjin launched new measures by setting up industries and the official banking firms came into being historically. Even today, the Tianjin locals still favor the Official Banking firm and resume the corresponding firms in the culture street.

官银号

官银号现位于古文化街宫北大街西侧，南邻商业广场。官银号曾是清政府官方设立的金融机构，1900年八国联军入侵天津之后，纷纷在天津设立银行，造成天津金融市场紊乱，导致国内钱庄银号萎缩。时任直隶总督兼北洋大臣的袁世凯在天津推行新政，命周学熙创建实业。北洋银元局、官银局、中国实业银号等金融机构应运而生，官银号很快发展成为实业发展服务的现代金融机构，促进了北方金融秩序的整顿。老天津人对官银号一往情深，于是就在古文化街恢复了"官银号"银行。

文化设施
Culture Facilities

Yuhuang Pavilion 玉皇阁

Yuhuang Pavilion is one of the oldest buildings and the largest Taoism temple in Tianjin. The original building group of Yuhuang Pavilion is very large, but it was seriously damaged when the imperialist intruded. The existing Qingxu Pavilion is the only Ming Dynasty building reserved.

玉皇阁始建于明朝洪武年间，是天津现存最古老的建筑之一，也是天津规模最大的道教庙宇。玉皇阁原建筑群十分庞大，由旗杆、牌楼、山门、钟鼓楼、前殿、八桂亭、清虚阁、南斗楼、北斗楼以及三清殿组成。后由于近代帝国主义的入侵，玉皇阁建筑群遭到了严重破坏，现存的清虚阁是保留下来的唯一一座明代建筑。

Tianhou Temple and Playhouse 天后宫和戏楼

The Tianhou Temple is located at the center area of Ancient Culture Street offering "Tianhou the Queen of Heaven" who saved lots of fishermen according to legend. There stand two big booms of 600 years history on Tianhou Temple Square. Tourists can see a memorial archway, and a bell tower and a drum tower at its two sides as almost they enter the front gate. And they can enjoy the scenic spots in the Tianhou Temple all the way.

The Playhouse is located at the east part of Tianhou Temple Square. It is an important cultural feature of the square. There are performances full of Tianjin characteristics such as comic dialogue, storytelling, Tianjin tricks and etc.

天后宫位于古文化街的中心地带，宫内供奉 "天后娘娘"。天后宫广场竖有两根大帆杆，帆杆已有600年历史。从山门进入天后宫，有"海门慈筏"牌坊一座，牌坊两侧分布着钟楼和鼓楼，其后是天后宫的前殿，前殿后依次为妈祖泉、正殿、药王殿、台湾殿、凤尾殿等建筑。

戏楼位于天后宫广场东侧，紧邻海河，是天后宫广场地带重要的文化点，日常演出的节目带有浓厚的津味特征。

Museum of Huaxia Shoe Culture 华夏鞋文化博物馆

This museum has six sections and fifty six specials including history of shoes, culture of shoes, boutique shoes and so on, with over one thousand shoes on display.

华夏鞋文化博物馆由百年老店老美华筹建，鞋文化博物馆中设有鞋履历史、鞋履文化、精品靴鞋等6个版块56个专题，1 000多件鞋展品。

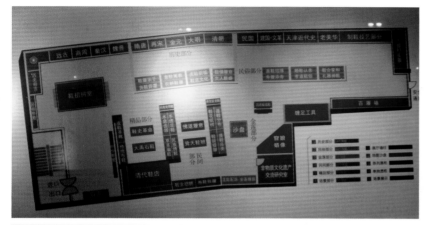

Imperial Song Street, Kaifeng
开封宋都御街

街区背景与定位
Street Background & Market Positioning

History 历史承袭

The Imperial Song Street is a cleared avenue at the south-north axis of city. It was a major road that emperors would pass when ancestor worship, southern suburb ceremony and touring were hold outside the palace. Therefore it was called the Imperial Street.

According to record, the street was 200 meters wide and was divided into three parts. The middle part was especially for imperial use and pedestrians were forbidden to enter into it. There were brooks full of lotus at the two sides, and peach trees, plum trees, pear trees and so on were at the two banks. Outside the brooks was civilian activity area where there were shops and was very bustling.

开封宋都御街是南北中轴线上的一条通关大道，它从皇宫宣德门起，向南经过里城朱雀门，直到外城南熏门止，长达十余里。它是皇帝祭祖、举行南郊大礼和出宫游幸往返经过的主要道路，所以称其为"御街"，也称御路、天街或者宋端礼街。

据孟元老的《东京梦华录》记载，御街宽约200米，分为3部分，中间为御道，是皇家专用的道路，行人不得进入。两边挖有河沟，河沟内种满了荷花，两岸种桃、李、梨、杏和椰树，河沟两岸有黑漆叉子为界。河沟以外的东西两侧都是御廊，是平民活动的区域。临街开店铺，老百姓买卖于其间，热闹非凡。每逢皇帝出游，老百姓聚在两边，争相观看皇家的尊严和气派。

Location 区位特征

Kaifeng is a famous city in the world in the 21st century. As early as Northern Song Dynasty, Kaifeng was a national political, economic and cultural center as well as one of the most prosperous cities in the world. There are many cultural relics from each dynasty and buildings of different styles of various dynasties. The newly built Imperial Song Street is constructed on its original site, extending from Xin Jiekou at the south to Wuchaomen at the north.

开封是21世纪国际上有名的城市，曾有"八荒争凑，万国咸通，集四海之奇珍"的辉煌历史。早在北宋时期，开封就是中国的政治、经济、文化中心，也是世界上最繁华的都市之一。各个朝代的更迭交替留下了众多的文物古迹，开封的仿古建筑群风格鲜明多样，宋、元、明、清、民国初期各个时期特色齐备。新建的御街在原御街遗址上修建，南起新街口，北至五朝门，全长400多米。

Market Positioning 市场定位

By optimizing environment, enhancing cultural connotation and fully using advantages of Kaifeng's tourist culture, architects try to build a Song Dynasty historic culture street and reproduce the prosperous scenery of the city.

通过工程的实施，优化古都环境，增强文化内涵，充分发挥开封市旅游文化的优势，打造"城市格局悠久、文物古迹丰富、古都风貌浓郁、北方水城独特"的宋代历史文化街区，实现"一城宋城半城水"的美好蓝图，再现其辉煌景象。

规划设计特色

Street Planning 街区规划

The planning area is the main body of Old Kaifeng. It includes "one street", Zhongshan Road, the axis of the old city; "one line", the line surrounding city wall; "two blocks", traditional blocks near Liujia Hutong and Shuanglong Lane; "four zones", Longting Lake Scenic Zone, Baogong Lake Scenic Zone, Fan Pagoda and King Yu Terrace Scenic Zone and "田" shaped central zone. It also contains three culture clues: "the blue", water system of Kaifeng; "the green", greenbelt surround city wall; "the purple", a commercial and trade zone centralizing Gulou Street.

The planning is aimed at six breakthroughs:

First, to repair the ancient city wall, build greenbelts inside and outside the protection zone of the city wall, and open up a tourist route along the city wall, at the same time to improve the greening rate of the city and tourist facilities.

Second, to increase water area and manifest the feature of northern water town through construction of water system.

Third, to build two inhabitant protection tourist zones of Liujia Hutong and Shuanglong Lane.

Fourth, to construct Song Gate, Cao Gate and Danan Gate Wengcheng Sqaure to fully show the grand and spectacular appearance of the ancient city wall of Kaifeng.

Fifth, to form a "田" shaped trade zone and speed up remolding constructions.

Six, to carry out the "town inside town" project development of Sansheng Street.

Generally, by putting the project to effect, to built the old city in to a distinct "Song Capital Scenic Zone" integrating tourism, shopping and holiday.

规划范围是开封古城主体，即现有城墙范围12.9平方千米，以及大南门以南至陇海线的部分外城区域约2.3平方千米。通过对"一街""一线""两片""四区"的修缮、保护、适度开发，彰显"蓝色""绿色""紫色"3条文化链。"一街"是古城的中轴线中山路；"一线"即环城墙一线；"两片"即刘家胡同、双龙巷附近的传统街区；"四区"即龙亭湖风景区、包公湖风景区、繁塔禹王台风景区和"田字块"中心区；"蓝色"是开封水系，"绿色"是环城墙绿化带，"紫色"是以鼓楼街为中心的金融商贸区。

规划旨在实现以下六大突破。

一是修复古城墙，建设城墙保护范围内外的绿化带、城北森林公园、孙李唐庄文化休闲区，开辟环城墙旅游线；建设西北湖公园、阳光湖公园、繁塔景区和法院街以北的旅游度假区，增加城市的绿地量和旅游设施。

二是通过水系建设，增加水域面积，彰显北方水城特色。

三是建设刘家胡同和双龙巷两个中原地区居民保护旅游区，开辟民居、民风、民俗游。

四是建设宋门、曹门门楼和大南门瓮城广场，充分体现开封古城墙巍峨壮观的风貌，提升古城形象。

五是形成"田字块"商贸区，加快山货店街、振河商业城西部、古楼商贸城和天中大酒店东侧等地块的改造建设。

六是实施三胜街周边集旅游、休闲、居住、购物为一体的"城中城"项目开发。

通过项目的实施，把老城区建成融旅游、购物、休闲、度假为一体，特色鲜明的"宋都景区"，实现与郑州的错位对接。

Street Design Features 街区设计特色

The design of Imperial Song Street highlights features of Song Dynasty. All shops of different sizes are built in Song style of two - or three-floor buildings higher at south and lower at north along the street. All the buildings are built of bricks and tiles and are decorated with red columns, painted beams and carved with Song windows. Major buildings are decorated with colored paintings of Song style.

Buildings, which apply Song style cross-shaped ridges and gable roofs, stand one by one. Uniformed east and west turrets at two sides of the memorial archway have ingenious and gorgeous appearances which are stunning. Two stone sculptures in front of the archway add dignity and imperial reverence. Words on horizontal inscribed boards, couplets, flags are all from historical record of Song Dynasty. More than 50 shops and stores are distinctive from each other. Salesclerks in Song-style costumes serve customers hospitably.

宋都御街在设计上突出宋代特色。以《东京梦华录》为蓝本，大小店铺均采用宋代营造方式，全街南高北低，多为二三层建筑，既有三步两店的一般店铺，又有体量宏大的仿古楼阁。沿街建筑青砖灰瓦，褐柱画梁，朱栏雕窗，主要建筑还施以宋式彩绘，使古都开封成为旅游开放城市中颇具魅力的一站。

楼阁相接，均采用宋代十字脊、歇山造等形式。乌头柱穿顶而出、朱柱矗立的15米高大牌坊，更为街区增添了诗情画意。牌坊两侧一式两幢的东西角楼，造型奇巧瑰丽，翼角振翅欲飞，姿态动人。牌坊前两侧两蹲石雕上骑着大象武士，更平添几分威严和皇权的敬畏。两侧角楼对称而立，楼阁店铺鳞次栉比，其匾额、楹联、幌子、字号均取自宋史记载，古色古香。50余家店铺各具特色，经营开封特产、传统商品、古玩字画。售货员身着仿宋古装，殷勤地招待八方来客，整个设计重现了宋都昔日的繁华景象。

Modern Imitated Commercial Streets of Ancient Style 现代仿建的商业街区

Major Commercial Activities 主要商业业态

Imperial Song Street is prosperous with so many shops. Most of the shop buildings are decorated with red bricks and grey tiles with colored flag flying in front of their doors and plaques dazzle visitors' eyes.

After years' rise and decline, many time-honored shops with profound cultural and historical connotations are kept to today. Now these shops of Song-style buildings display commodities of the 1990s and their operators are modern youths who know business management very well and full make use of advantages of location to let visitors experience folk customs of Song Dynasty when sightseeing and shopping. Especially during the Song Culture Festival in April every year, many service staffs in ancient costumes serve customers politely, making visitors feel a distinctive fun when walking in the street.

宋都御街大小店铺林立，买卖兴隆，一派繁荣景象，有药店、茶社、酒楼、宾馆、旅社、银行、医院、镖局、画院、照相馆、糖烟酒、百货楼、美术楼等，这些店铺大都是红砖灰瓦，门前彩旗飞舞，纱灯高照，牌匾夺目，在蓝天白云的衬托下，更显得不凡。

几经兴衰，老字号仍然留存至今。如"又一新"、"第一楼"、"陆稿荐"、"晋阳豫"、"包耀记"、"马豫兴"、"大金台"、"万福楼"、"王大昌"、"乐仁堂"、"老五福"等，具有深厚的文化和历史底蕴。如今，在这些具有宋代建筑风格的店铺里，陈列的是20世纪90年代的商品，而店铺的经营者也大多是具有竞争意识的当代青年。他们深谙经营之道，充分利用地利特点，把新潮的公司、商场名字改称为具有宋代特色的招牌，像"惠民药局"、"东京镖局"等，使人们在游览购物的同时，充分领略宋代都会的民风民俗。特别是每年四月的宋都文化节期间，许多店铺服务员身着古装，以礼待客，置身其间，更有一番悠思别趣。

Featured Commercial Area 特色商贸区

Fanlou Curio Town

According to historical records, Fanlou Curio Town is the place where Emperor Huizong of Song Dynasty had date with Li Shishi, a famous prostitute. It has classic and grand builded style. Located at the southwest corner of Longting Square, it receives millions of tourists every year.

矾楼古玩城

矾楼古玩城是历史记载中宋徽宗和李师师约会的地方，建筑风格古典宏大，位于AAAA级景区龙亭广场西南角，景区每年接待中外游客超百万，是古城旅游中心。位于龙亭广场角上的古玩城是开封市规模最大的古玩交易市场。

Food & Beverage 餐饮类

Daoxiangju

Daoxiangju was opened in 1882. It sold fried dumplings, wontons and other small dishes with wine in the beginning. Later it added fried dishes to become a special flavor restaurant.

稻香居

"稻香居"原名"天津稻香居锅贴铺",开业于清光绪八年(1882年),原以经营锅贴、馄饨为主,兼营下酒小菜,后增添炒菜,成为著名风味小吃馆。店铺原在马道街北口,后迁入宋都御街。

Cross-bridge Rice Noodle Restaurant

Cross-bridge rice noodle is a special food with more than one hundred years history in Yunnan.

过桥米线馆

过桥米线是云南滇南地区特有的食品,已有百余年历史。此种食品主辅合一,深受各类人群的喜爱。过桥米线馆以特有的滋嫩、鲜香、清爽适口、富于营养而著称。

Leisure 休闲类

Dasong Wine Club

Depending on the research result of Song wine and adopting mode of combining sightseeing and producing, wine-making technology of Song Dynasty and Chinese wine culture are wholly reproduced to build the first wine museum themed by Song Dynasty wine culture in central plain of China. The decoration of Dasong Wine Club continues the style of Song Dynasty. With lighting effect, the decoration highlights the wine in the club incisively and vividly.

大宋酒道馆

开封大宋酒业是专注于大宋系列酒品研发、生产、销售的综合性企业。公司位于七朝古都开封万岁山游览区，占地面积30余亩，是国内少有的花园式酒类企业。项目建设依托宋酒研究成果，按照宋酒历史秘方，运用宋代酿制程序，采取了旅游观光与生产展示相结合的模式，全景再现了宋代的酿酒工艺及中国酒类文化的博大精深，建成了中原地区首座以宋代酒文化为主题的酒类博物馆。大宋酒道馆的装修沿用了宋朝文化的风格，加上灯光的照射效果，将馆内的酒品演绎得淋漓尽致。

Curios 古玩类

Heyuxuan

It sells ancient porcelains, bronze wares, ancient jade articles, ancient furniture, celebrities' paintings and calligraphies, badges and so on.

和韵轩

和韵轩主营古瓷器、青铜器、古玉器、古漆器、古代家具、古印石、金银器、名人书画、徽章、杂件类等。

Kaibao Hall

Kaibao Hall is an art gallery of Kaifeng embroidery. Its main products are Kaifeng embroideries and Suzhou embroideries. It also sells celebrities' paintings and calligraphies, treasured swords, jades, bracelets, and male kung fu outfits, embroidered high-heels, flat satin shoes and so on.

开宝堂

开封简称"汴",故开封地域的手工刺绣被称为"汴绣"。开封在历史上是北宋都城,刺绣已达鼎盛,汴绣在继承宋朝手工制绣的基础上发展起来,因而汴绣亦称"宋绣"。开宝堂汴绣礼品艺术馆位于龙亭公园前宋都御街57号,主营汴绣、苏绣,兼营名人字画、镇宅宝剑、翡翠玉器、手镯,另有丝绸睡衣、睡裙、肚兜及男式功夫衫、绣花高跟鞋、平跟绸缎鞋等。

Zangwangge

It specializes in selling collections of jade articles, paintings and calligraphies, curios and so on.

藏王阁

藏王阁是一家专门经营玉器、字画、古玩等收藏品的店铺。

Gutiange

It is a professional old shop selling celebrities' paintings and calligraphies, scholars' four jewels and high class art books.

古天阁

古天阁是一家集名家字画、文房四宝和高档书画类图书于一体的专业老店。

Boyazhai

Its main products include jade articles of official ware, old coins, New Year paintings, Kaifeng embroideries, calligraphies, copies of ancient books and celebrities' letters, and so on.

博雅斋

博雅斋主营官窑玉器、钱币年画、汴绣精品、书法、油画、版画、古籍碑拓以及名人手札。

Wutongxuan

It sells celebrities' paintings and calligraphies, the scholar's four jewels, seal cuttings, art tools, art books and so on.

无同轩

无同轩主营名家字画、文房四宝、湖笔名砚、苏娟湖绫、金石篆刻、字画装裱、美术画材、艺术图书等。

Xibu Alley, Zhangjiajie
张家界溪布街

街区背景与定位 / Street Background & Market Positioning

History 历史承袭

Xibu Alley is a project of a huge investment by Zhangjiajie Jinrui Tourism Development Co., Ltd.
Xibu is a dialect. It means an exquisite handmade brocade of Tujia minority and is a royal tribute. Xibu Alley implies a street with stream and a place full of joys.

溪布街是由张家界金瑞旅游发展有限公司斥巨资开发的项目，项目占地面积约百亩，总建筑面积约7万平方米，位于张家界武陵源区。
溪布是土家语，又名西兰卡普，是一种精美的土家族手工织锦，是土司王进献给皇室的贡品。溪布街的寓意是布满小溪的街道，充满快乐的地方。

Location 区位特征

Xibu Alley is located at the core of Wulingyuan Scenic Reserve in Zhangjiajie. It is facing Zhangjiajie Theater at the east, adjacent to Baofeng Bridge at the west, next to Wuling Avenue at the north and Suoxi River at the south. Therefore, it has a strong tourist appeal. The project has a convenient traffic. It is only about 30,000 m away from Zhangjiajie city center and is located at the self-drive toursim circle of Changzhutan, Yiyang, Loudi and Changde.

溪布街位于张家界最为著名的武陵源景区核心，街区东临张家界大剧院，西连宝峰桥，北靠武陵大道，南依400米长的索溪河河岸水景，具有极强的旅游号召力。周边交通便利，距离张家界市中心仅30余千米，位于长株潭、益阳、娄底、常德3小时自驾旅游圈内。

Market Positioning 市场定位

Xibu Alley depending on customs and charm of Tujia and Miao minority and integrating fashionable elements is forged to be the only one comprehensive commercial pedestrian street gathering waterside bar street, Chinese famous snack street, folk shopping boutique street and so on.

溪布街街区以土风苗韵为基调，融合符合时代审美的时尚休闲元素，全力打造成为一条集水上酒吧街、中华名特小吃街、湘西民俗购物精品街、休闲客栈、创意工坊等为一体的复合型旅游商业步行街，成为武陵源核心景区不可复制的文化旅游藏品。

Street Planning 街区规划

1. Block General Planning

It is organically arranged in a general courtyard layout by using combination of building shape and size, and following the linear drift theory. Buildings are romantic and elegant to show required fashion and simplicity of featured commercial street, and meanwhile to satisfy the functional and aesthetic demands.

Its main entrance is at the east near Zhangjiajie Theater. The whole building cluster is orderly the entrance zone at the east of Baixi Gulley, the middle leisure zone and terminal urban business zone at the west, and the riverside landscape area along Suoxi River.

2. Function Division

The whole block is generally divided into three commercial modes. They are shopping street, restaurant street and leisure culture street. All these functional streets are complement to each other.

1. 街区总体规划

整体规划院落围合式布局，利用建筑形体和体量的组合，以线性飘积理论为原则，进行有机的布置；建筑力求浪漫、飘逸和通透，体现旅游特色商业街所具备的时尚和拙朴的气质，同时满足现代商业街区的功能和审美需求。

旅游商业街区的主入口设在用地的东面，毗邻张家界大剧院，以现有广场共同组织为入口广场，整个建筑组团依据三区一带的原则，依次为：百溪沟以东的入口区、百溪沟以西的中部休闲区和末端的城市商业区、沿着索溪河形成河滨休闲景观带，以打造靓丽的滨河景观。

2. 功能分区

整个规划片区按照业态划分，大体上共分为3种类型的业态模式，即购物街区、餐饮街区及休闲文化街区。各个功能街区之间力求做到相互渗透，并以同业差异、异业互补为原则，进行相对的功能片区划分及业态的有机组合。

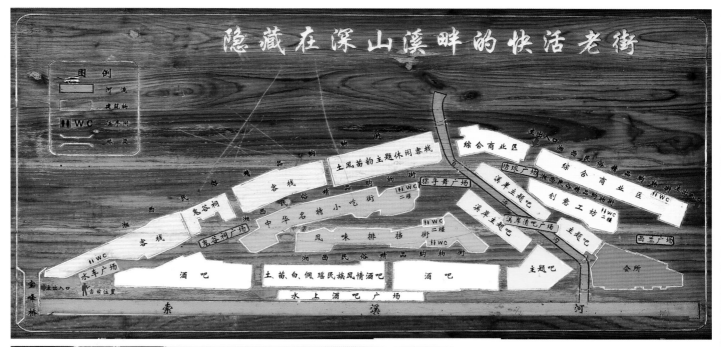

Street Design Features 街区设计特色

Redundancies are left as far as possible from streets to courtyard to enrich the shape of building space and experience. Traditional languages of Tujia and Miao minority are applied in façade design after art reprocessing.

The design of plants fully considers the soil, season change and color scheme to create different sceneries in different seasons. Meanwhile a stable and natural plant community is formed to provide visitors with relaxed and pleasant feeling.

项目建设采用丰富的空间语言进行建筑的平面功能设计，从街道、入户至庭院都尽可能地留有一定的空间冗余量，丰富建筑空间的形态和体验感。设计师在立面造型上将土家族和苗族的传统语言通过艺术再加工和简洁的抽象变形，作为建筑表现的构成元素，使立面风格统一，古朴大气。

植物设计力求虚实有序，通过树影、声响、叶色等传递风、月、云、日、气、四季等自然信息。在植物配植上，充分考虑武陵源地区的土壤特点、植物四季季相更替和色彩搭配，使其在不同的季节形成不同的景致，同时形成稳定、自然的生态植物群落，在视觉上给游客传递轻松和愉悦的感受，形成"以人为本"的空间环境。

Major Commercial Activities 主要商业业态

商业业态 Commercial Activities

Commercial activities of Xibu Alley are composed of three parts — restaurant, shopping and leisure. Big restaurants are arranged at entrance zone and along Wuling Road; tourist shops are at two sides of interior streets; and leisure culture shops are along Baixi Gulley. The middle leisure zone at the west of Baixi Gulley is equipped with posts, urban supporting commercial comprehensive street, tourist souvenir street, local specialty street, leisure culture street and so on.

溪布街的商业主要以餐饮、购物和休闲三大部分组成：入口区部分由入口广场及沿武陵路设置的大型餐饮店、内部街道两旁设置的旅游购物店、沿百溪沟一侧设置的休闲文化店组成；百溪沟以西的中部休闲区由上至下分别是驿站和城市配套性商业综合街区、旅游购物店街区、餐饮小吃及地方特色街区、旅游购物街区、休闲文化街区组成；末端城市商业区主要分布购物、餐饮等。

Featured Commercial Area 特色商贸区

Attention was paid to the commercial division at the planning stage of Xibu Alley. The division is as follow.

溪布街在规划之时就注意商业的分区，具体而言，其大体商贸区如下。

Waterside Bar Square

Waterside Bar Square is located at the bank of Suoxi River. There are mainly soft rocks bars, hot dance bars and perform bars. Each bar is equipped with courtyards, operating rooms and rest rooms to provide owner with convenient operating environment except waterside bars.

水上酒吧广场

水上酒吧广场位于索溪河沿岸，该区域主要以慢摇、热舞、演艺酒吧为主，总面积约8 000平方米，一般酒吧面积为120~180平方米，以南北朝向为主。该区域集合全国知名的旅游酒吧和名店名企。除水上酒吧商铺外，该物业每一间都带有庭院、操作间及卫生间，为经营者提供便利的经营环境。

Western Hunan Folk Boutique Shopping Street

In addition to Baixi Gulley Bar Area and Suoxi River Bar Area, the first floors of all the other streets are boutique shops and the second floors are with a few handicraft workshops. The house types of the shops are designed in square without columns inside. It not only collects brand-name and high-quality local specialties of western hunan, but also assembles tourist boutique shops, featured handicraft workshops and so on, therefore, it has become the first choice for tourists when traveling and shopping in Zhangjiajie.

湘西民俗精品购物街

从溪布街主入口开始向绣球广场方向延伸，除百溪沟酒吧区和索溪河酒吧区外，其余街区一楼都是购物精品街商铺，局部有少数特色店及手工艺坊在二楼。购物街商铺总面积达20 000平方米，商铺面积大都为在25~35平方米。商铺户型设计方正，铺面内无柱体，购物街总商铺达800家，购物街不仅汇集了湘西的名优土特产，更集合了来自全国各地知名旅游目的地的精品购物店、特色工坊、特色旅游纪念品，成为游客在张家界旅游购物、逛街的首选目的地。

China Famous and Featured Snack Street

This street is located at the second floor of Xibu Alley's central area. All the shops are square with completed function. All the shops are equipped with smoke exhaust pipes, and they can be equipped with kitchen and operating room when running. Famous snacks all over the country are gathered here to let visitors have tastes of Chinese food.

中华名特小吃街

中华名特小吃街位于溪布街中心区域二楼，总面积约10 000平方米，其中名特小吃街约3 500平方米，商铺面积一般为25~35平方米，约120间铺面，铺面方正，功能齐全，每一间商铺都设有排烟管道，在经营时可在店内设置厨房及操作间。中华名特小吃街将汇聚湘西、湖南以及国内如四川、广东、重庆等地的知名小吃，让游客在溪布街就可领略到中华美食的魅力。

Comprehensive Commercial Street

Comprehensive Commercial Street is next to Xiuqiu Square and Creativeness Studio. There are mainly shops of national features selling handicrafts.

综合商业街

综合商业街临近绣球广场、创意工坊，商业街内店铺以卖手工工艺品为主，多是具有民族特色的商铺，所卖的商品有土家族的西兰卡普、苗族的银饰等。

Theme Square

There are four theme squares at Xibu Alley. They are Waterwheel Square, Xilan Square, Xiuqiu Square and Waving Dance Square.

Waterwheel Square is located at the main entrance beside Baofeng Bridge. Themed by waterwheel, it creates a countryside rural atmosphere which provides visitors with a physical and psychological quietness.

Xilan Square is located at the entrance near to Creativeness Studio. According to its theme of brocade of Tujia, there is a wooden board at the center introducing the brocade of Tujian and wall paintings showing the working process to highlight the preciousness of the national craft.

Xiuqiu Square is beside Western Hunan Folk Boutique Shopping Street and facing Xilan Square. It is themed by a hanged vermeil ball made of strips of silk.

Waving Dance Square is built according to requirement of waving dance. Wooden pillars with flags and lanterns are set all around the square. Leather drums are set at two sides and bonfire area is at the middle.

主题广场

溪布街区中共有四大主题广场，分别是水车广场、西兰广场、绣球广场和摆手舞广场。各个广场各有特色，与所在的功能区主题一致。

水车广场位于宝峰桥侧的主入口处，靠近索溪河，以水车建筑为主题，营造出乡村田园的氛围，让游客刚进入街区便仿若置身于田园生活之中，还身体与心灵的一份宁静。

西兰广场位于创意工坊、会所一侧的主入口处，以西兰卡普为主题而设立，广场中设置木牌匾专门介绍西兰卡普，广场外侧的墙壁上有展示西兰卡普制作过程的壁画，凸显出民族工艺品的珍贵与精美。

绣球广场位于湘西民俗精品购物街的一侧，与西兰广场相对而望，广场以一个悬挂着的朱红色绣球为主题。

摆手舞广场按照摆手舞的要求而建，场四周设置幡旗和挂有灯笼的木柱，场地的边侧设有牛皮大鼓，场地中央辟为篝火区。

水车广场

西兰广场

绣球广场

摆手舞广场

Food & Beverage 餐饮类

Zuixian Lou

Zuixian Lou is a restaurant for receiving large tour groups. Its dishes are mainly Hunan cuisines made of indigenous materials.

醉仙楼

　　醉仙楼是接待大型旅游团的餐厅，位于湘西铺子上层，最多可接纳700余人就餐。餐厅内菜品以湘菜为主，口味清爽，菜的选料都是本地原产的绿色食品。

Jia'niang

Jia'niang is a shop that specializes in selling home-brews. Its interior is decorated with gourd drinking vessels, knives and swords, therefore customers can feel a bold and generous atmosphere.

家酿

　　家酿是一家专营自家酿制的美酒的商铺，里面酒的品种众多，既有独具特色的敲敲酒（又名湘西土家竹筒酒），又有普通的包谷酒、糯米酒，还有烧刀子、女儿红、土司王酒、压寨夫人酒等。店铺内部多用葫芦形盛酒器品修饰，还摆有刀剑等器物，颇具豪迈古风。

Tea Story

Tea Story engages in creating brand new beverage form. It integrates traditional milk tea with healthy tea culture to offer healthier special drink.

Tea Story has the delicate decoration. Its chairs and tables are simple yet fashionable and the book shelves against walls add relaxing atmosphere. Outside its windows is the wonderful scene of Xibu Alley and Baofeng Lake.

茶物语奶茶

茶物语隶属于湖南商贸有限公司，是一家主打绿色健康休闲饮品的特权经营机构，致力于打造全新的饮品形态，将传统奶茶与健康茶文化相融，提供更健康的特色饮品。

茶物语奶茶（溪布街店）装修别致，室内座椅简约而时尚，倚墙而靠的书柜平添了几分轻松的气氛，窗外是美丽的溪布街景色，通过窗子可以远眺宝峰湖的美景。

Xiangxi Puzi

Xiangxi Puzi is located at the prime section of Xibu Alley. There are exhibition areas of wax printing, silverwares, stone paintings, Hunan embroidery and so on.

湘西铺子

湘西铺子位于张家界溪布街黄金地段，内设大湘西蜡染、银器、石头画、湘绣等展示区，各种湘西纯手工艺品琳琅满目，极具民族地域特色。

Villas 会馆类

Guanshanyue Villa

It is a holiday villa in quadrangle courtyard layout of creative garden style. It enjoys a advantageous location which is overlooking Suoxi River, facing Baizhang Valley and far seeking the Tianzi Mountain. There are 35 suites of rooms, dinning area, nature auditorium, air garden café, chess and card room and so on.

观山悦公馆

观山悦公馆是一家四合院格局、创意性、园林式的度假公馆，地理位置优越，可俯瞰索溪河，平视百丈峡，远眺天子山。公馆内拥有各种特色客房35套，还有茶饮就餐区、自然大讲堂、空中花园咖啡屋、棋牌室等配套设施。

BaLing Villa

The main body of BaLing Villa is a brick and wood structure building. There are wooden chairs and tables in the outdoor. Its interior furnishings are common wares in peasant family of Western Hunan. It is simple yet charming and is a good place for leisure.

捌零会馆

捌零会馆主体是一座砖木混合结构的建筑，室外设有露天的木质座椅，内部用湘西农家常用的器具装饰，显得朴素而又别具韵味，是休闲的好去处。

Bars 酒吧类

Blues Bar

Blues Bar is right beside Baixi Gulley with a stone bridge linking two banks in front of its door. Its door is just like a big cask, and its window is hanged with bottles in the way of hanging maizes. They are really striking.

布鲁斯酒吧

布鲁斯酒吧倚百溪沟而建，门口有一座石桥连接两岸。酒吧入口形似一个大酒桶，窗口挂满了酒瓶，与农家丰收后悬挂玉米的方式极为相似，十分夺人眼球。

Yesterday Bar

It has two floors. Its door is extremely stunning, for its shape is like a strip of film embracing the door. There are pictures of famous singers in old days on the film, which are antique yet fashionable, and highlights the name of the bar. There is a wooden stair outside the door to the second floor.

昨日重现酒吧

昨日重现酒吧高两层，门口的装修异常醒目，一条类似胶卷的装饰建筑环绕酒吧大门，胶卷中印有旧日著名歌星的图片，复古而时尚，也凸显了酒吧的店名。门口设有木质楼梯，可直通二楼。

Cultural Facilities 文化设施

Guigu Temple 鬼谷祠

Guigu Temple is a building of Qin and Han dynasties style. It has ten exhibition rooms. It is an important window showing national spirit and folk culture of western Hunan. Its first floor mainly introduces Guigu master and his followers, and the second floor introduces how Guigu master influenced the culture of Western Hunan.

鬼谷祠是一座秦汉风格的建筑，青瓦红柱，雕梁画栋，共两层，十个展厅，面积将近1 000平方米。它是展示湘西民族精神、民俗文化与湘西文化脉络的重要窗口。第一层展厅以介绍鬼谷子及其知名弟子、春秋战国文化为主；第二层展厅介绍鬼谷子对湘西文化的影响。

Characteristic Memorial Archway 特色牌坊

There are many memorial archways at Xibu Alley. The shapes of the archways reveal a strong impression of Tujia and Miao. It has a decorating effect as well as a guidance function.

溪布街内设有许多的牌坊，牌坊的造型体现出浓浓的土苗风情，既起到装饰的效果，又有着引导指示的作用，是溪布街内绝妙的一景。

引导指示系统
Guidance & Sign System

Lingnan Impression, Guangzhou
广州岭南印象园

Street Background & Market Positioning
街区背景与定位

History 历史承袭

Lingnan Impression is located at the south part of Guangzhou Higher Education Mega Center HEMC where originally the Lianxi Village is. Lianxi Village was called the Guangzhou HEMC Museum. In 2005, because of the construction of HEMC, the village was removed entirely and its historic value came to light. Now 11 buildings including ancestral halls, oyster shell walls and so on are preserved after repairing. Besides, five residential buildings were immigrated in and 38 unique style buildings were built.

With traditional crafts, food and arts of Lingnan features, Lingnan Impression is a window showing intangible cultural heritage of Lingnan.

岭南印象园位于广州大学城南部，原练溪村区域，总占地面积16.5公顷。练溪村曾被称为广州大学城博物馆，2005年，为配合小谷围的大学城建设，整体拆迁，其历史价值再度被发现。现在，原练溪村里的11栋建筑经过修复，保存下来，包括宗祠、蚝壳墙等，另外从村外移来了5栋民居建筑，同时新建了38栋风情建筑。

岭南印象园景区注入了岭南地区传统的民间工艺、饮食文化、表演艺术等，是广绣、广彩、粤剧、木偶戏等岭南特色艺术、工艺的一个集中展示平台，成为承载岭南文化的载体，展示岭南非物质文化遗产的窗口。

Location 区位特征

It is located at Xiaoguwei Island surrounded by water and facing the Pearl River, therefore, it has an excellent scenic view. With annular and radiated road network on the island and 16 bus routes connecting downtown, it has a very convenient traffic. There are many scenic spots around Lingnan Impression, including Chimelong Tourist Resort, Baomo Garden and so on.

岭南印象园位于广州大学城所在的小谷围岛，四面环水，面朝珠江，景观视线极佳。周边交通十分便利，岛内已建成环形加发射的道路网络，有16条与市区相连的公交线路。岭南印象园周边景区众多，如长隆旅游度假区，番禺区内的宝墨园、余荫山房等，并与这些旅游景区形成联动发展。

Market Positioning 市场定位

Lingnan Impression depending on its advantageous location and profound folk customs creates a large-scale leisure cultural tourist attraction integrating cultural sightseeing, interesting entertainment, featured food, leisure shopping and splendid shows.

岭南印象园依托天然的地域优势和原练溪村浓厚的岭南民风民俗文化，建立了一个融文化观光、趣味游乐、特色餐饮、休闲购物、精彩演艺于一体的大型休闲文化旅游景区。

Street Planning 街区规划

Lingnan Impression includes two function zones — Lingnan Culture Sightseeing Zone and Mountain Top Leisure Park. Lingnan Culture Sightseeing Zone is organized by Lianxi Street features traditional residences, local expression and folk customs. Mountain Top Leisure Park is designed to satisfy visitors' demands for getting close to nature, appreciating nature and so on.

岭南印象园包括岭南文化游览区和山顶休闲公园两大功能区。岭南文化游览区以岭南传统民居、岭南乡土风情和岭南民俗文化为主，整个游览区由练溪大街组织。山顶休闲公园可满足游客亲近自然、欣赏自然风景、野外休闲的需求，以景区正门北侧的山林为主。

规划设计特色
Planning & Design Features

Street Design Features 街区设计特色

Lingnan Impression is a typical Lingnan traditional building group. The residences are built along the river. Long Qingyun Lanes, antique security doors, delicate Manchu windows, twisted stream, and clear pond diffuse a charm of Lingnan water village. There are also nostalgic places such as old wine shop, old barbershop, old cinema, old photo studio and so on, and folk craft exhibition zones such as Xinhui Palm-leaf Fan House, Guangzhou Embroidery House and so on.

岭南印象园是典型的岭南传统风格建筑群落。园内民居依水而建，或窄门高屋，或镬耳高墙。悠长的青云巷、古朴的趟栊门、精致的满洲窗，小溪蜿蜒，池塘清澈，散发出岭南水乡的韵味。园内有包拯相祠、敬佛堂、天后宫、华光古庙等民间祭拜文化区，又有老酒坊、老理发店、老电影院、老照相馆等令人怀旧的场所，还有新会葵艺馆、广绣馆、木雕宫灯等民族工艺展示区，让游客时刻感受到浓浓的岭南风情。

Modern Imitated Commercial Streets of Ancient Style
现代仿建的商业街区

Major Commercial Activities 主要商业业态

The commercial activities of Lingnan Impression are mainly food, shopping and shows.

As for food, there are Guangdong featured snacks of different flavors, such as Nanxin Stewed Milk Beancurd, Shanshui Beancurd Jelly, Shunde Fried Ice and so on.

As for shopping, there are so many crafts, such as products of Palm-leaf Fan House, Guangzhou embroideries, wooden carved palace lanterns and so on.

As for shows, there are lion and dragon dances, dragon-boat race, Huaguang memorial ceremony. Besides, there are also interactive performances of traditional wedding ceremony, shadow figures show, marionette show and so on.

　　岭南印象园是集文化观光、趣味游戏、特色餐饮、休闲购物、精彩演艺于一体的大型休闲文化旅游景区。区内商业以餐饮、休闲购物和节目演出为主。

　　餐饮方面，景区安排了不同口味的广东特色小食，如南信双皮奶、山水豆腐花、顺德炒冰、酸辣汤等，岭南特色手信也是一应俱全，有鸡仔饼、客家酿酒、小磨香油等。

　　休闲购物方面，景区内有众多的工艺品，如葵艺馆编制的产品、广彩广绣、木雕宫灯、西关木屐等。

　　节目演出方面，岭南印象园内囊括了舞狮舞龙、菠萝鸡鸣、赛龙夺锦、祭华光帝等传统岭南文化节目。除此之外，还有还原的传统婚嫁仪式的互动式表演"绣球招亲"以及陆丰皮影、五华提线木偶、岭南杂耍等。

品牌商铺展示
Brand Shops

Food & Beverage 餐饮类

Daming Hall Stewed Milk Beancurd

Daming Hall is an old brand of Shunde stewed milk beancurd. Its shop of traditional Guangzhou building style is small and can house about ten people. It also provides customers with outdoor chairs and tables for rest.

大鸣堂双皮奶

大鸣堂是顺德双皮奶的老字号，店面不大，可容纳十余人。店面是传统的广州风格建筑，店门外放有露天的座椅供食客休息。双皮奶营养丰富，能促进脑细胞发育，调整血气，对骨骼生长有极佳的作用，是适宜儿童成长、中老年人抗衰老、女性养颜的佳品。

Lianxi Hotel

Lianxi Hotel is opposite Lingnan Theater. It is a three-floor Lingnan featured cultural building. Its dishes are mainly Cantonese cuisine and Hunan cuisine.

练溪酒家

练溪酒家位于岭南剧场对面，楼高3层，是具有岭南特色的文化建筑。酒店的主要菜式是粤菜和湘菜，招牌菜有烤鸡、清蒸皖鱼、蒸猪肉丸等。

Lianxi Teahouse

The interior decoration of Lianxi Teahouse is very distinctive. It is equipped with chairs and tables made of bamboo. Against its wall is a row of bamboo shelves placing tea bread, which can not only embellish the space, but also sell tea. There are big red lanterns on the wall which provide warm and amiable feelings.

练溪茶楼

练溪茶楼内部装修很有特点，室内是清一色的竹椅、竹桌，靠墙是一排放着茶饼的竹架子，既用来装饰空间，又售卖茶叶。墙上还挂有大红灯笼，在灯光的映衬下，茶楼显得红红火火，让顾客感觉到热情的氛围。茶楼内主要供应面饭、粽子、豆腐花、甜品等小吃，物美价廉。

Tingzai Porridge

Tingzai Porridge is a famous Guangzhou local snack from Liwan Xiguan Area. It was sold at small boat ("Tingzai" in Cantonese) on Litchi Bay River by some fishermen's families, therefore it was called Tingzai Porridge.

艇仔粥

艇仔粥是广州的著名小吃，最早来自广州荔湾西关一带，原为一些水上人家用小艇（广州话叫"艇仔"）在荔枝湾河面经营贩售，故名"荔湾艇仔粥"。艇仔粥以新鲜的小虾、鱼片、葱花、蛋丝、花生仁、海蜇、浮皮、油条屑为原料，依照滚粥冲烫粥料的手法，立即冲滚，稍后就能品尝。其特点是粥底绵烂，粥味鲜甜，软滑兼备。

Grain Store

Grain Store is a store selling onsite made sesame oil, peanut oil, tofu pudding, soya-bean milk and noodle.

为民作坊

为民作坊是一家手工现场制作芝麻油、花生油、豆花、豆浆、挂面的店家，让游客既能体验制作美食的传统手艺，又能品尝天然健康的食品。

Shopping 休闲购物类

The Macao Story

The Macao Story is a shop retailing crafts. Its interior is decorated in old Macau style and presents a unique Macau flavor.

澳门故事

　　澳门故事是一家零卖工艺品的商铺，室内装修成老澳门的风格，体现出澳门独特的风情，让游客在游乐中购物，其乐无穷。

Woman Boudoir

Woman Boudoir is a store specializing in lady products in old days, such as perfumes and cosmetics, clothes and purses used by Xiguan ladies. It interior is furnished as boudoir of Xiguan Ladies'.

美女私房

　　美女私房是一家专营旧时女士用品的商店，里面售有古时西关小姐使用的胭脂水粉、服饰、包包之类的商品。室内布置成西关小姐闺房的样式，让游客在购物时感受到西关小姐的日常生活场景。

Others 其他商铺

70's Lane 70巷

The 70's Lane is decorated according to the theme of life scenes of the 1970s. It is furnished from aspects of daily used coal, daily food and

70巷是以广州20世纪70年代的生活场景为主题布置的,包括生活用品、日常食品和体育等多个方面,让游客穿越时空,仿佛身处20世纪70年代的广州生活场景之中。

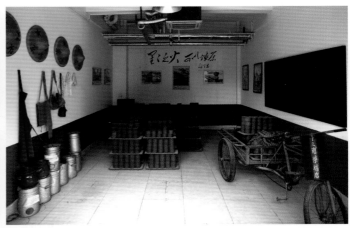

Red Image Museum 红色映像馆

It is at the Ancestral Temple of Huo's Family where visitors can watch old movies in the courtyard and review funs in childhood.

红色映像馆建在练溪霍公祠上,游客可在映像馆的露天庭院里观看老电影,重温儿时的乐趣。

Yiyan Ancestral Temple & Huaiai Ancestral Temple 婚庆堂&同庆堂

Yiyan Ancestral Temple was built in Qing Dynasty. Its main building is three rooms wide and two parts deep with pan tiles, sculptures, eave tiles and triangle-shaped edge. Its beam structure keeping the ancient wooden structure is simple. Huaiai Ancestral Temple is right beside Yiyan Ancestral Temple. They show the wedding custom of Lingnan together.

婚庆堂建于清代，是陆姓氏族的宗祠。主体建筑面阔三间，深二进，二进均为镬耳山墙。屋面布板瓦，灰塑瓦脊，素胎勾头、滴水。梁结构件简单，柱础保留有早期木构做法的"木质"。同庆堂与婚庆堂并排，共同展现出岭南的婚庆习俗。

Lianxi Opera Stage 练溪大戏台

Lianxi Opera Stage is an important place for shows, such as traditional operas and sketch shows, in Lingnan Impression.

练溪大戏台是岭南印象园内一个重要的节目演出点，游玩岭南印象园时，练溪大戏台不可错过，在这里可以看到岭南戏曲、小品等多种类的节目表演。

引导指示系统 Guidance & Sign System

Litchi Bay, Guangzhou
广州荔枝湾

街区背景与定位 / Street Background & Market Positioning

History 历史承袭

Lichi Bay is located at Pantang, the west part of Guangzhou. Pantang was a world of waters in ancient times. In Tang Dynasty by the impact of sediment of the Pearl River it became a land. Its terrain is flat and with lots of pool and low-lying land, therefore it was called "Bantang" which means that a half of it is composed of water. Its name was changed into "Pantang" during Qianlong Period. Zhengong Dike of Tang Dynasty, Liu's Hualin Garden in Southern Han Dynasty and former site of West Royal Garden are located here. The water and soil in Pantang were very fertile and produced many rich arrowheads, water caltrop, lotus roots, wild rice shoots and water chestnuts which were praised as five beauties in Pantang. In 1958, Liwan-Lake Park was built.

荔枝湾在广州城区西部泮塘一带，古时泮塘一带为一片汪洋，到唐代时由于珠江泥沙冲击而成陆地，地势低平，多为池塘、洼地。由于一半是池塘，因而人们约定俗成地称之为"半塘"。古时，人们称学宫为"泮宫"，入学宫读书成为"入泮"，为图吉祥，在清乾隆年间，"半塘"改称"泮塘"。唐朝时的郑公堤、南汉时刘氏的华林园就在泮塘地区，西御苑旧址也在此地。古时此地以种植茨菇、菱角、莲藕、茭笋、马蹄为主，因泮塘水土肥沃，这五物特别肥美，被誉为"泮塘五秀"。但随着城市发展，大量农田被用作住宅和工厂，泮塘五约曾一度只保留了一些砖木建构的低矮民房。1958年，政府号召群众义务劳动，在泮塘内开挖荔湾湖，后建成荔湾湖公园。2007年，广州市人民政府又清拆泮塘五约内民房，腾出3.6万平方米土地并入荔湾湖公园。

Location 区位特征

Litchi Bay is located at Longjin West Road, Liwan District. The Pantang Road in this area is a road of most Xiguan style. Pantang Road is 390 meters long, along which are tourism spots like Liwan-Lake Park and Renwei Temple Square, and famous restaurants like Panxi Restaurant and Meishiyuan.

荔枝湾位于广州市荔湾区龙津西路、荔湾湖公园以及泮塘五约一带，区内有最具西关风情的道路——泮塘路。泮塘路长390米，沿线分布着荔湖公园、仁威庙广场等旅游景观和泮溪酒家、美食园等品尝佳肴的去处。另外还有蒋光鼐故居、小红楼、风水基民宅等旅游景点沿荔枝湾涌两岸分布，旅游资源丰富。

Market Positioning 市场定位

In the comprehensive improvement of residential environment of Liwan District in 2010, Litchi Bay was positioned as a landmark tourism area possessing profound culture and gathering food, culture and leisure tourism.

在2010年荔湾区开展的人居环境综合整治中，荔枝湾被定位为具有浓厚文化及风情，且集美食、文化、休闲旅游于一体的标志性特色旅游区。

规划设计特色 Planning & Design Features

Street Planning 街区规划

The improvement of Litchi Bay takes protecting traditional culture as principle. It controls the landscape environment, heights of buildings and style of buildings at surrounding region. Therefore, it should protect individual buildings meanwhile protect the general traditional scene of this area including architectural space form, residential culture and commercial culture. In addition to protecting remains of history, protecting old famous shops, traditional craft and brand is also very significant to the protection of tourism resource of unique local feature.

荔枝湾的整治以保护传统文化和地区特色的延续性为原则，在划定的保护范围内，控制周围地区的景观环境、建筑高度和建筑风格，在重视"点"状建筑保护的同时，重视"面"状历史地段的整体街区传统风貌的保护，包括保护视线景观环境、建筑空间形式、居住文化及商业文化等。比如，在上西关涌段，将其功能区划分为"商埠广场"、"民俗社区"和"文化走廊"，在恩宁路大地涌段节点划定"食"、"艺"、"器"、"货"等区域。另外，在尽量保护历史的真实遗存，特别是对区内的文物保护单位和特色建筑应遵守不改变原状的前提下，对商业"老字号"的保护重点是保护其传统字号、传统工艺、传统建筑和驰名品牌，保护地方特色浓郁的旅游资源。

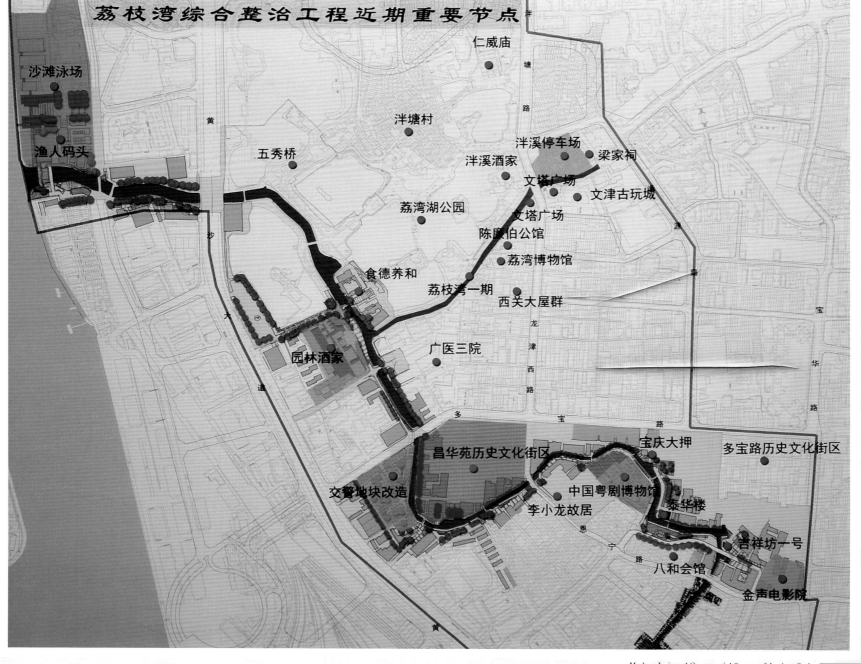

Street Design Features 街区设计特色

Pantang has become a bustling commercial area with strong folk customs in Guangzhou since Ming Dynasty and there are a group of streets and buildings of strong Lingnan features. Streets and buildings built in different periods of different styles reflect the history and culture of Guangzhou in different developing stages to some extent, and create a simple and elegant city style of profound Lingnan culture.

In 2010, the reconstructed Pantang Road keeps its Xiguan style with traditional style buildings.

荔湾区泮塘在明代就已成为广州商贸繁荣和民俗风情较浓的地区，历史上形成了一批极具岭南特色的街区和建筑群，其中有石脚水磨青砖的西关大屋、优雅的花园式别墅、名人公寓、中西合璧的骑楼等。这些不同时期、不同风格的街区和建筑在一定程度上反映了广州各个发展时期的历史和文化轨迹，形成了古朴典雅、中西交融、岭南文化气息浓郁的城市风貌。

2010年经过改造后的荔湾区仍坚持"西关风格"，泮塘路沿线两侧建筑物立面的涂料以灰、白色为主，天花采用青灰色琉璃瓦，墙面使用仿古青砖，招牌及墙柱则使用饰面花格，并有传统中式纹样，体现出浓郁的岭南建筑风格。另外新建有广州美食园，以彰显"食在广州，味在荔湾"的独特魅力。

Modern Riverside Scene at Qingming Festival
现代清明上河图

Major Commercial Activities 主要商业业态

As the saying goes, "eating in Guangzhou, tasting in Liwan", there are many delicacies in Litchi Bay area such as Panxi Restaurant, Xiguan Shijia Restaurant and other old famous restaurants or snack shops. Beside, based on the profound history and culture, the Wenjin Antique Town was built.

荔枝湾是集美食、文化、休闲旅游于一体的特色旅游区。素有"食在广州，味在荔湾"之称的荔枝湾少不了美食，在饮食业中，除了有名气响亮的泮溪酒家和西关世家两大园林酒家，荔湾区还建有广州美食园，引进多家老字号企业，如利口福、致美斋、莲香楼等，打造老字号手信街。除此之外，荔湾区泮塘还依托自身深厚的历史文化，打造文津古玩城，发展古玩市场。

Featured Commercial Area 特色商贸区

In order to give full play to the commercial potential of Pantang, the government plans featured commercial areas as follow.

为充分发挥荔湾区泮塘地区的商业潜力,政府还规划了以下特色商贸区。

Wenjin Antique Town

Wenjin Antique Town, which is reconstructed from grocery company Longjincang, is a project developed by Guangzhou supply and marketing cooperative and Liwan District Liyuan state-owned assets management Co., ltd. It is the largest modern professional curio mall in Guangzhou with over 360 shops. Over 400 m² modern convention center is preserved on the fifth floor to aperiodically hold curio art exhibition and auction.

文津古玩城

文津古玩城由原来的日杂公司龙津仓全面改造、装修而来,是广州市供销合作总社与荔湾区荔源国有资产经营管理有限公司共同合作开发的项目,经营面积近10 000平方米,拥有360多家商铺,是广州规模最大、集交易和展示服务于一体的现代化专业古玩商城。古玩城以经营古旧瓷器、字画、家具、玉器等为主,开设了现代名家臧品专区与国家级大师工作室。古玩城五楼预留有400多平方米的现代会议展示中心,不定期地举办古玩艺术品展会、拍卖会等。

Xiguan Mansion Protection Zone

There are many traditional residential buildings of Xiguan from the late Qing Dynasty to early Republic of China in the Xiguan Mansion Protection Zone. Their plan layouts, façades, sections and decorations are of profound local features.

西关大屋保护区

西关大屋保护区的具体范围南至三连直街，东到龙津西路，西临西关上支涌，北接逢源沙地一巷。区内多清末至民国初期西关的传统民居建筑，其建筑的平面布局、立面构成、剖面到西部装饰等都有着浓厚的广州地方特色。

Modern Imitated Commercial Streets of Ancient Style
现代仿建的商业街区

品牌商铺展示 Brand Shops

Food & Beverage 餐饮类

Panxi Restaurant

Panxi Restaurant was founded by Li Wenlun in 1947. At the beginning it was a small restaurant built of bamboo and pine skin on a pond. Now it is a national super restaurant. The restaurant was designed by Mo Bozhi, a famous landscape architect in China.

The layout of Panxi Restaurant is well-arranged. With Liwan Lake beside it, the restaurant is enjoying a wonderful scene. The exterior of the restaurant is poetic, and its interior is twisting with rich space layers. The landscape made by fine workmanship is striking. Precious interior decoration, furniture and adornment are stunning.

In addition to Cantonese dishes, Panxi Restaurant also provides dishes of Sichuan, Shandong, Huaiyang, Chaozhou and Asia-Pacific dishes.

泮溪酒家

泮溪酒家于1947年由李文伦创建，创建之初只是一家用竹木、松皮搭架在荷塘之上的充满乡野风情的小酒家，以酒家附近的一条小溪命名。现在，泮溪酒家已是国家特级酒家、中华餐饮名店。它位于广州西郊荔湾湖畔，是广州三大园林酒家之一，酒家由中国著名园林建筑师莫伯治设计而成。

泮溪酒家由假山鱼池、曲廊、湖心半岛餐厅、海鲜舫等组成，布局错落有致，加上荔湾湖景色衬托，更显得四处景色如画。酒家外围粉墙黛瓦，绿榕掩映，大门上墨绿色的牌匾由朱光市长所题。酒家院内布局迂回曲折，空间层次丰富，几组园林的布置也显得十分睿智。如迎宾楼下面的假山布置是依据"东坡游赤壁"的图谱而建成的，巨构精工，与曲桥流水相映成趣。而假山上的迎宾楼博取我国四大名园之精华，飞檐翘角，四面以五彩花窗装嵌，十分清雅瑰丽。楼内更是艺术的殿堂，金碧辉煌的木雕檐楣、泛金套色的花窗尺画，都是珍品。厅堂内琳琅满目的酸枝家具、名人字画让客人叹为观止。

泮溪酒家除了粤菜外，还有川、鲁、淮扬、潮州和亚太（地区）菜等菜式，另外还有广州市人民政府命名为"名点"的八大美点，这些使泮溪酒家成为了集饮食于一体的"美食园"。

Xiguan Shijia

Xiguan Shijia is located at the shore of beautiful Liwan Lake and next to Renwei ancestor temple of thousand years history. Classic beauty is the most striking characteristic. All the details of its building reveal the Xiguan feeling of old days. Its major dishes are Cantonese cuisine.

西关世家

西关世家园林酒家位于风景秀丽的荔枝湾荔湾湖畔，与有千年历史的仁威祖庙毗邻。古典美是酒家建筑的最大特色，古朴的大门、木趟栊的栅门、花雕大窗、木质宫灯、精致屏风、满洲窗、青砖绿瓦、翘角飞檐、民族壁画、粤剧名伶无一不透出昔日西关风情。其主要菜式是粤菜。

Crafts 工艺类

Canton Enamel and Embroidery

Canton Enamel and Embroidery are featured craft in Guangzhou. Canton enamel features tight composition of pictures, dark and bright color, and splendid and magnificent characteristic. With over 300 years history from Ming Dynasty to Qianlong Period of Qing Dynasty, Canton enamel gradually became a unique artistic style. Canton embroidery is one of the four top embroideries in the country. Canton embroidery can be classified into two kinds. One is dazzling and dignified, and the other is colorful and exquisite.

The art gallery of Canton Enamel and Embroidery is located near the Literary Tower. Its arrangement is very simple yet charming with gorgeous Canton enamel placed at the left and exquisite Canton embroidery at the right.

一彩一绣

"一彩一绣"是广州地区的特色工艺文化，"一彩一绣"即"广彩"、"广绣"。广彩是广州地区釉上彩瓷艺术的简称，以构图紧密、色彩浓艳、金碧辉煌为特色，犹如万缕金丝织白玉，又叫广州织金彩瓷，始于明代的广州三彩，到清代发展为五彩，并在乾隆年间逐步形成独特的艺术风格，有300多年历史。广绣是全国四大名绣之一，专指广府地区的刺绣工艺品，包括刺绣字画、刺绣戏服、珠绣等。广绣大致分为两大品类：一是盘金刺绣，二是丝绒刺绣。盘金刺绣以金线为主，辅以彩纷刺绣，金碧辉煌，灿烂夺目，雍容华贵。丝绒刺绣开丝纤细，色彩缤纷，绣出的花鸟尤其精美。

何丽芬、梁秀玲一彩一绣艺术馆位于荔湾区文塔附近，室内左边放置有精美华丽的广彩作品，右边是精细的广绣作品，布置简单，却韵味十足。

Former Residence of Jiang Guangding 蒋光鼎故居

Jiang Guangding is a famous anti-Japanese high-ranking military officer. His former residence was built in the late Qing Dynasty and early Republic of China. The building is a courtyard residence of Lingnan architectural style. It has both characteristics of Xiguan Mansion and Western culture.

蒋光鼎先生是我国著名的抗日将领、原国民革命军第十九路军总指挥，其故居建于清末民初，建筑为岭南建筑风格的内庭式住宅，面积766平方米，既有传统西关大屋的风格，又有西洋文化的特点。

Liwan Museum 荔湾博物馆

Liwan Museum, which was founded in 1996, is the first district-level museum in Guangdong. It mainly collects, exhibits and studies the history, culture and customs of Liwan. The museum used to be the residence of Chen Lianzhong, the comprador of HSBC in the period of Republic of China. It is composed of Chinese courtyard and Western villa.

荔湾博物馆成立于1996年,是以收藏、陈列和研究荔湾历史、文化、民俗为主要内容的广东省首家区级博物馆。博物馆原为民国时期汇丰银行买办陈廉仲先生故居,为民初的中式庭园、西式别墅,庭园面积1 000余平方米,有仿罗马、希腊的柱式和拱门,又有由峰峦、岩洞、亭台、路桥等组成的奇景。

Renwei Ancestor Temple 仁威祖庙

Renwei Ancestor Temple was built in 1052 and was repaired and expanded during Ming and Qing dynasties. The dignified and antique temple is full of distinct features of Lingnan traditional crafts. The wood carvings of lotus, lions and so on are very exquisite and present the graceful and magnificent characteristics of wood carving art in Qing Dynasty.

  仁威祖庙始建于北宋皇祐四年（1052年），经明清两代维修扩建后，广三路深五进，占地面积2 200余平方米。庙宇建筑庄严古朴，富有鲜明的岭南传统工艺特色。大殿内木雕艺术品比比皆是，如大殿梁枋的托脚雕成莲花和倒挂鳌鱼形状，梁枋驼峰雕刻成两对瑞狮，线条丰富流畅，狮子的须毛镌刻细致。木雕采用了圆雕、通雕、浮雕、沉雕的手法，又敷设金彩，流光横溢，体现出清代木雕艺术雍容华丽的特色。

Literary Tower 文塔

Literary Tower is located at a small courtyard named Yunjinyuan opposing Panxi Restaurant. The tower is 13 meters high, with a hexagon pinnacle wood and brick structure. The tower is enshrining and worshipping Wenquxing, the star of wisdom, therefore it was a lucky tower of scholars in ancient times.

文塔位于泮溪酒家对面的小庭院"云津苑"内,高13米,为六角形金字尖顶砖木结构,塔底每面宽2.5米,塔高两层,塔门向北。塔基座为石砌,塔身为东莞大青砖所砌,以陶瓷葫芦为塔顶,高达2米。塔内供奉文曲星,是旧时文人心中的福贵塔。

Xiguan Mansion 西关大屋

Xiguan Mansion is a traditional residential form with distinct Lingnan characteristics in Guangzhou built by rich merchants in Liwan District during the late Qing Dynasty. Xiguan Mansion is mostly built of brick and wood and its layout is in accordance with traditional house of Central Plains. In their courtyards there are plants, rockery and fishpond. The whole environment is quiet and beautiful.

　　西关大屋是清末豪门富商在广州城西荔湾区一带兴建的极具岭南特色的广州传统民间住宅形式，多为砖木结构，青砖石脚，高大的正门用花岗石装嵌。其平面布局按中原传统的正堂屋形式，典型平面为三间两廊，左右对称，中间为主要厅堂。中轴线由前而后，由南而北，依次为门廊、门厅（门官厅）、轿厅（茶厅）、正厅（大厅或神厅）、头房（长辈房）、天井、二厅（饭厅）、二房（尾房）。每厅为一进，一般大屋为二三进，形成颇多的中轴线。庭园中栽种花木，筑有假山鱼池，颇为典雅清幽。

The Red Building 小红楼

The Red Building which is located at No.33 of Fengyuan Street was built in the period of Republic of China. It was designed by a German designer. Its decorations are delicate and exquisite. It is a typical Western style building in modern times.

逢源大街33号旧民居小红楼建于民国时期，为四层砖混结构，由德国设计师设计，形式典雅，装饰细致精美，是较典型的近代西式风格民居建筑。

Small Painted Boat Studio 小画舫斋

Small Painted Boat Studio was built in 1902. It is a ring form garden style Xiguan Mansion with houses all around and a garden in the middle. The building is delicate and exquisite with profound Lingnan architecture charm.

小画舫斋建于清光绪壬寅年（1902年），是一座环形园林式西关大屋，四周为楼房，中间是花园，楼房精致典雅，具有浓郁的岭南建筑韵味。其整座建筑以白花岗石脚、水磨"东莞青砖"精砌墙壁，平滑洁亮。

Wenfeng Gujie, Chongqing
重庆文峰古街

街区背景与定位
Street Background & Market Positioning

History 历史承袭

The predecessor of Wenfeng Gujie is Baita Street because of the pure white Wenfeng Pagoda. The place was a former business center as well as the transit depot of Sichuan salt, but later depressed due to the decline of local piers. In 2010, Jintiandi Real Estate Group spent two years to restore this historical site. The program launched on May 1, 2012.

文峰古街的前身为"白塔街",因有一座主体白色的文峰塔(清嘉庆十五年修建)而得名。这里曾经商贾云集,也是"川盐"中转地,后来码头没落,白塔街因此萧条。2010年,锦天地产集团用时两年修建文峰古街,并于2012年5月1日开街。

Location 区位特征

Wenfeng Gujie which centers on Wenfeng Pagoda, local well-known landmark building, is located at riverside region of Hechuan District, the confluence of Jialing River and Fujiang River. It spans to the Nanping Bridge under construction to the east, and borders the Fujiang Bridge on the west. It leans against the Bingjiang Road to the north and overlooks the three rivers to the north. It connects to the Hechuan North City Commercial Center Bridge as well as the national scenic spot, Diaoyu Castle Bride and is backed by the Hechuan administration center as well as the southern new city. Its prominent geographical position enjoys very convenient traffic.

项目以合川家喻户晓的地标性建筑——文峰塔为核心,地处合川区南屏片区北部滨江地带,嘉陵江与涪江汇流之处,东至在建的嘉陵江南屏大桥,西邻涪江一桥,南倚滨江路,北眺三江。古街与合川北城商业中心一桥相通,与国家级风景名胜区——钓鱼城一桥相连,背倚合川行政中心区和南部新城,区位优越,交通便利。

Market Positioning 市场定位

Wenfeng Gujie gathering commercial streets, luxury residence areas as well as riverside ecological park is a comprehensive development project. With traditional commercial streets as major business, the project models on Hongya Cave, local famous landmark, to develop a commercial street which integrates tourism, dining, leisure, culture and entertainment.

文峰古街是集仿古商业街、高档住宅区、滨江生态公园于一体的综合性开发项目。商业部分为仿古传统商业街,以重庆著名地标"洪崖洞"为蓝本,打造合川集旅游、餐饮、休闲、文化、娱乐于一体的巴俗文化商业街区。

Street Planning 街区规划

Planning & Design Features 规划设计特色

The project is located in the core area of Nanping area, covering about 250,000 m² of land. The greening rate is 39%, and the whole plan can be described as "one street, one park and two areas."

"One street" refers to the traditional commercial street which starts from the Fujiang Bridge in the west and stops at the salt warehouse in the east. The street that lasts about 1,000 m is core of the area. It is a commercial street which gathers tourism, dining, leisure, culture and entertainment.

"One park" means the riverside ecological park which makes the most of the original gully topography to develop water system as well as beautiful ecological wetland. It blends the green of the park organically into the urban environment to improve regional greening level and bring nice landscape. Famous Hechuan ancient buildings of Tang and Song dynasties are also restored, and combined with classical Chinese gardening techniques, so as to reflect the culture and history of the old city and improve the historic and cultural value of the park. The riverside ecological park can not only improve the program image, but also serve as the green lung of the southern city. It will carry on the city history and improve the space quality of the whole city.

"Two areas" stand for the east and west residential areas in the south of the program. Both areas rely on the appeal of the commercial street and riverside ecological park on the spot to develop tourism commercial properties and drive property development effectively, so as to realize the commercial and cultural value of the residential programs. With excellent river scenery and good landscape environment, those residential areas are regarded as the most cultural and environmental-friendly high grade residential areas in Hechuan District. There are about 1,200 households at local, and the permanent residents are approaching 4,000.

　　项目位于南屏片区规划的重点核心区域，总建筑面积约250 000平方米，绿地率为39%，整体规划分为"一街、一园、两片"。
　　"一街"将开发建设仿古传统商业街，西起涪江一桥，东至盐仓，是项目规划的核心区域，长约1 000米，是集旅游、餐饮、休闲、文化、娱乐于一体的巴俗文化商业街区。
　　"一园"即滨江生态文化公园，充分利用原冲沟地形，开发河湖水系，打造优美的生态湿地，将公园绿化有机地融入城市环境中，提升区域绿化水平和景观均好性。恢复唐宋合州著名古建，如岁寒亭、清华楼、凌霄阁、荔枝阁、濂溪祠，结合中国古典造园手法，反映合州历代文化历史风情，提升公园历史文化价值。滨江生态文化公园不仅可以提升项目的形象，同时也是合川南城片区的绿肺，将承载起时代的追溯、历史的变迁，提升了整个城市的空间品位。
　　"两片"即项目南侧东西两片住宅区，东侧为联排别墅、多层花园洋房、小高层住宅、高档居住小区，西侧为多层住宅、小高层住宅、中高档住宅小区，借助景区商业街和滨江生态文化公园的吸引力，发展旅游商业地产，有效带动物业开发，实现住区商业文化价值。两片居住区既有良好的观江视野，又有良好的自身景观环境，从而成为合川区最具文化及环境品位的高档居住社区。"两片"住宅区共计约1 200户，常住人口接近4 000人。

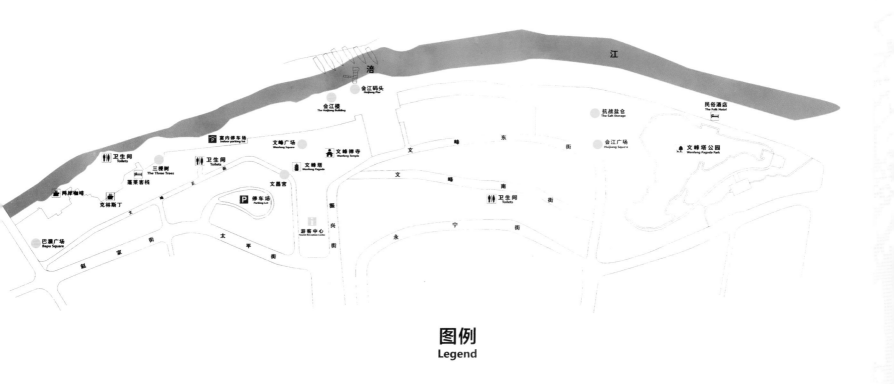

图例
Legend

Street Design Features 街区设计特色

Exquisite traditional architectural details and numerous iconic architectural symbols are used to restore the city memory as well as its historical context, so as to reproduce the authentic life of Bayu area, showing a historical scroll of Bayu culture.

In terms of architectural style, the program combines tenon-and-mortise wall in eastern Sichuan folk houses, stilted building particular in Chongqing, and architectural style of commercial buildings and classical official building of ancient China. Warm colors are taken as main hue.

For architectural layout, traditional geomatic theory is fully applied to arrange all buildings with an orderly hierarchy. Corridors, down channels, riverside walkways constitute a multi-dimensional circulation system accessible to any places within the program.

As for commercial planning, four theme functional areas are arranged to build an integrated commercial economy chain covering shopping, leisure, dining and entertainment.

文峰古街以精致的传统建筑细节与众多突出的标志性建筑符号，还原城市记忆与历史文脉，从而再现巴渝原味生活，呈现出一幅巴渝文化的历史长卷。

在建筑风格上，文峰古街结合了川东民居的穿斗墙、重庆特有的吊脚楼、古代商业建筑与古典官式建筑风格，色彩以暖色系为主，整体上具有亲切热闹的亲民化风格，具备观光旅游的特色商业优势。

在建筑布局上，文峰古街充分运用中国传统风水学，整体布局错落有致、进退有序，以回廊、下行通道、临江步行栈道构建起了立体多维的交通体系和人流体系，使整个商业街区可以随意贯穿、自由通达，既符合古代街市内聚人气的传统特色，又承袭现代开放式步行街集聚人流的时尚商业业态。

在业态规划上，文峰古街以四大主题功能区，构建一体化商业经济链，用多主题街区组成购物、休闲、娱乐、餐饮互为补充的多元商业业态。

Major Commercial Activities 主要商业业态

Leisure: landscape tea house, tea house, café, featured inn, and spa club.
Recreation: KTV, club, featured bar, traditional wine shop, disco bar, cinema and water games.
Food & Beverage: Chinese restaurant, hot pot, private cuisine, and special snacks.
Shopping: local products, crafts, nonmaterial cultural products.

休闲：观景茶楼、茶艺馆、咖啡厅、特色休闲、特色客栈、水疗会所、三江垂钓。
娱乐：KTV、会所、特色风情酒吧、传统酒肆、风情迪吧、电影院、水上娱乐。
餐饮：大型品牌中餐、大型品牌火锅、精品私家菜、各地方风情餐厅、特色小吃。
特色购物：地方土特产、地方手工艺品、地方非物质文化产品等。

商业业态
Commercial Activities

品牌商铺展示
Brand Shops

Food & Beverage 餐饮类

He Jia Sheng Yan
With stilted building of Ming Dynasty as prototype, the restaurant reflects the unique architectural style and profound historical context of Chongqing.

和家盛宴
该店以重庆明清时代的吊脚楼为原型，采用分层筑台、退台以及吊脚楼架空的形式，天井、花台与数重院房错落相间，小青瓦、石板、木雕门窗等集中体现了独特的重庆建筑风貌和厚重的历史文化。

Veggie Restaurant
Veggie Restaurant, just as its name implies, mainly caters for vegetable dishes and imperial cuisines.

素餐厅
素餐厅以普通素菜和仿膳为主，从调料到食材的选择都非常讲究。酱油、醋等调味品皆选用传统方法自然酿制的，大多数食材来自于原生态，在保证纯天然的基础上，加入素食净心、养生的理念，开发出了多款营养搭配、有利于身心健康的素食菜品，以此达到"茹素净心，打开灵性之门"的境界。

Yan Gui Lai

Yan Gui Lai covers dining and accommodation. The architectural layout includes front and rear yards and combines with natural landscapes. The building reflects the features of Bayu garden and folk house at mountainous area.

燕归来

燕归来集餐饮、住宿为一体。建筑布局采用了前院后园的形式，与自然山水结合，展示了巴渝园林以及山地民居的特色。

Recreation 休闲、娱乐类

Kristine Club

Kristine Club is a comprehensive recreational place for coffee, foot massage, Western cuisines and KTV. The archaic environment as well as elaborate dishes will break through geographical boundaries and bring urban tourists extraordinary feelings.

克林斯丁

　　克林斯丁是融合了咖啡、足艺、西餐、KTV于一体的综合性娱乐休闲场所。这里古朴的环境，精心制作的各色美食，突破地域界限，为都市游客带来不寻常的感受。

Ye Shang Music Pub
Ye Shang Music Pub is positioned as a theme music pub of healthy entertainment and fashion life. It brings an entertainment revolution to the night life at the ancient street.

夜尚酒吧

夜尚酒吧是定位于健康娱乐和时尚生活的主题音乐酒吧，为古街夜生活带来了一场革命性的娱乐风尚。

Commercial Pedestrian Street, Beichuan
北川商业步行街

Street Background & Market Positioning
街区背景与定位

History 历史承袭

According to history record, Beichuan is the born place of Yu the Great, one of the ancestors of Chinese nation, the founder of Chinese first class society slave society and a hero of preventing floods. So far, there has kept lots of historical relics about Yu the Great. It is a place of interest containing cultural landscapes and natural landscape.

Long history and reclusive living environment enable Beichuan completely to keep its traditional culture of Qiang, such as buildings of antique beauty, bright and colorful clothes, unique customs, mysterious belief and worship, and create a splendid grand spectacle of Qiang's folk custom.

据史籍记载，北川是中华民族的人文初祖之一、中国第一个阶级社会——奴隶社会王朝的创立者、治水英雄大禹的降生之地。唐代以前，县境就建有众多的大禹庙，每年农历六月初六大禹诞辰举行祭祀活动的民间习俗延续至今。至今，仍保存着大量有关大禹的历史遗迹，是集人文景观和自然景观于一体的大禹故里风景名胜，已成为人们访古探幽的圣地。

悠久的历史与长期闭塞的生活环境，使北川完整地保存了羌族传统文化：古色古香的羌族建筑、绚丽多彩的羌族服饰、独具一格的羌族习俗、神秘的信仰崇拜，构成了一幅壮美的民俗画卷。

Location 区位特征

It is located at the space axis of Beichuan New County. It is next to Yuwang Bridge at the west and adjacent to Anti-seismic Memorial Park and Beichuan Culture Center at the east. It is only 20 kilometers away from Mianyang city center.

商业街位于北川新县城城市空间轴线上，西接禹王桥，东接抗震纪念园和北川文化中心，距离绵阳市中心仅20千米，成都市民驾车仅1.5小时即可到达。

Market Positioning 市场定位

Themed by national customs and featuring folk crafts, the Commercial Pedestrian Street is a comprehensive tourist commercial pedestrian street of shopping, food, entertainment, leisure and relaxation.

项目定位为以民族风情为主题，以民族民俗手工艺展示为特色，集购物、餐饮、娱乐、休闲、休憩等功能为一体的综合型旅游文化商业步行街。

Street Planning 街区规划

规划设计特色
Planning & Design Features

The Commercial Pedestrian Street is one of the ten symbolic projects that Shandong Province provides assistance in construction of Beichuan New County. The planning is integrated with ideas of urban design. By highlighting local feature, optimizing layout of mountains and rivers and creating comfortable size, it manifests the new appearance of Beichuan. Traditional architectural symbols, decorative patterns, modern materials and techniques are organically combined together, and local plants, folk activities and public arts and traditional crafts are inherited to create the building style of new Beichuan and unique city features.

At the same time, the outlines of buildings should give way to outlines of mountains. Taking example of space arrangement of Qiang race village, four or five units are combined to a building with courtyard, instead of one building for one household. The idea of structure adjustment of block space layout centralizes Beichun New County to promote the construction of industrial park, and accelerates the change of development mode of Beichuan. In addition, geography condition, traffic condition and existing development basis should be taken into consideration to satisfy living demands of residents. What's more, it should weaken the function of town development axis and emphasize passageway function of life line in tourist area.

北川商业步行街是山东省援建北川新县城的10大标志性项目之一，是北川新县城景观中轴线和步行廊道的重要组成部分，也是集中体现传统羌族风貌特色的重点区域。规划融入城市设计的思想和方法，通过突出地域特色、优化山水格局和塑造宜人尺度等3个方面，彰显北川新的风貌特色。通过川西民居和羌寨等传统建筑符号和装饰图案与现代建筑材料和施工技术有机融合，塑造新北川的建筑风格；结合乡土树种的栽植、民俗活动的展示、公共艺术的传播以及羌族手工业的继承等塑造城市特色风貌。

同时，该项目坚持建筑轮廓线要让位于山的轮廓线，借鉴羌寨的空间聚落手法，将四五家单位合起来建一座楼，形成一个院落，而不是一家一栋。街区空间布局结构调整的思路首先以北川新县城为中心，结合安昌镇是县域人口、产业发展的核心地区，促进对口支援产业园区的建设，加快北川发展模式的改变，在原北川的山区不再鼓励进行大的产业项目的布局。其次，对街区的广大地区，综合考虑地质条件、交通状况、现有发展基础等因素，重点扶持其中综合条件较好的街巷，提高街区抗震防灾能力，增强社会公共服务提供能力，扩大公共服务覆盖面，满足区域内乡镇居民的需求。再次，弱化城镇发展轴线的作用，突出旅游区域生命线通道功能，以"一心、多点、多廊道"组织空间结构。

Street Design Features 街区设计特色

The basic design conception is from Qiang race village. The choice of materials, expression of forms and creating of environment are rooted in local culture and feature of Qiang race.

The combination of antique charm and modern elements makes residential buildings of Qiang irradiate. Every detail of the buildings is dotted with traditional Qiang race symbol. And all the buildings are connecting with each other by corridors, overpasses, and stairs; therefore the buildings are independent yet communicate with each other.

The Commercial Pedestrian Street is in frame structure. Most buildings are three-floor and their flat roofs and slope roofs are interspersed with each other. Façade adopt cultured stone, raw wood instead of Al-alloy doors and windows. Colors of buildings are sandy beige interspersed with brown of woods. Walls at the side of the street are embedded with sheep totems.

Traditional watch tower of the Qiang race is used for several times in the design. Now the watch towers are for traffic and landscape function, rather than defense function. There are three watch towers with elevator and viewing deck in the whole street. After the New County is built, visitors and local people can appreciate the appearance of the whole Beichuan from these watch towers.

项目设计的基本构思源于羌寨聚落。材料的选择、形式的表现和环境的营造都根植于羌族的地域文化与本土特色，是四川西北地区最大的特色文化旅游商业街区和步行街。

古风遗韵与现代元素相融合的设计使历经千百年的羌族民居建筑焕发出新的光彩。建筑的每一处檐角、女儿墙顶、门窗都点缀着传统的羌族民俗文化符号——白色羊皮褂、白头帕、白布裙……这些建筑彼此之间通过走廊、天桥和楼梯连接，既独立又连通，自成单元又家家相通。

步行街为框架结构，各楼多为3层，屋顶采用平坡结合并相互穿插，整体看上去错落有致。立面采用文化石、原木等传统建筑材料，避免了铝合金门窗。建筑色彩采用片石浅褐色点缀原木的棕褐色，街区两边建筑的墙上嵌入羊图腾，街头矗立羌式牌坊，朴素自然，充分展现了羌族特色。

在设计中，多次运用碉楼这一传统羌族元素。碉楼本身的战时防御功能已不复存在，取而代之的是交通和景观作用。整个商业步行共有3座碉楼，最高的碉楼为37.7米，其他两座碉楼约25米。37.7米高的碉楼是整个北川新县城的最高建筑，碉楼设有电梯、观景台。新县城建成后，前来观光的游客、当地民众可以登上碉楼俯瞰整个北川新县城的风貌，感受北川羌族特有的建筑风貌。

Major Commercial Activities 主要商业业态

The commercial activities of the street include five types — featured retails, fashionable retails, featured restaurants, leisure entertainments and service supports. Featured retails include tourist souvenirs, local specialties, crafts and arts; fashionable retails include creative products, travel products, outdoor goods and fashionable life supplies; featured restaurants include local special restaurants, business restaurants, leisure restaurants, featured snacks and so on; leisure entertainments include theme teahouses, theme cafés, culture exhibitions, private museums and so on; and service supports include theme hotels and inks.

　　商业街业态主要涵盖特色零售、时尚零售、特色餐饮、休闲娱乐、服务配套5大类型。特色零售包括旅游纪念品、土特产品、工艺品、艺术品等；时尚零售包括创意产品、旅游用品、户外用品、动感时尚生活用品等；特色餐饮包括当地特色餐饮、商务餐饮、休闲餐饮、特色小吃等；休闲娱乐包括主题茶坊、主题咖啡、中式保健、文化展示、个人专题博物馆等；服务配套包括主题宾馆、客栈等。具体包括有重庆渝富侨、丽江客栈、哎呀呀饰品、成都伤心凉粉、成都高记白家肥肠粉、上海休闲会所、台湾连锁"一点味"快餐、绵阳天香悦酒店、绵阳海上海酒店、绵阳四海香中餐馆、北川禹露茶叶、北川阳光尔玛绣品、北川羌山绿野食品、北川幸福羌绣、北川古羌水磨漆、安县明珠旅游工艺品、江油肥肠等商家。

Featured Commercial Area 特色商贸区

Yuwang Bridge

Yuwang Bridge is a covered bridge over Anchang River, connecting Anbei Road at the west and adjacent to Qiang Ethnic Commercial Street. It is a three-hole reinforced concrete arch bridge of 204.2 meters long. Its middle is walkway and its two sides are for shops. On the bridge are stone - and wood - made buildings presenting Qiang ethnic impression and Yu the Great Culture, and at the two ends of the bridge are traditional Qiang watchtowers.

禹王桥

禹王桥是一座风雨廊桥，由山东省对口援建，横跨安昌河，西接安北公路，东连羌族特色商业街。桥型为三孔钢筋混凝土连拱桥，采用混凝土箱型断面，桥梁全长204.2米，中间为走道，两侧为预留店铺，桥面面积2 700平方米。桥面上为体现羌族风情、大禹文化的石木结合的建筑。桥两端为传统的羌族碉楼。

引导指示系统

Guidance & Sign System

China Qiangcheng, Wenchuan
汶川中国羌城

街区背景与定位 / Street Background & Market Positioning

History 历史承袭

Qiangcheng Weizhou Town located at the northeast part of Wenchuan County. Its population is mainly Qiang ethnic minority and some Zang, Hui and Han people. This area is full of ravines and its weather is dry. There is a Buwa local residential group of yellow soil, a national cultural relic unit and some other tourist resources, and rich mineral resources. On May 12th, 2008, this town was seriously damaged in Wenchuan Great Earthquake.

羌城威州镇位于汶川县境内东北部，镇域面积134平方千米。全镇辖12个村，常住人口3万左右，镇内人口以羌族为主，兼有藏族、回族和汉族。镇区内沟壑纵横，岭峦错落，气候干燥，日照丰富，植被稀疏。辖区内有国家级文物单位布瓦黄泥土碉群及姜维城、七盘沟、茨里高山水库等旅游资源和汉白玉、石英石、金刚砂、硫铁矿等矿产资源。2008年5月12号，汶川特大地震使该镇遭受严重损坏。

Location 区位特征

Weizhou is attached to Wenchuan County and is 146 kilometers away from Chengdu. It is passed through by 213 and 317 national road. It is the traffic, political, economic and culture center of Wenchuan County. It possesses rich mineral resources and tourist resources.

威州镇隶属于汶川县，是汶川县城所在地，距离四川省省会成都市146千米，境内有213、317国道穿过，是汶川县的交通、政治、经济、文化中心，也是进出阿坝州的交通要塞。镇内矿产资源和旅游资源丰富，有姜维古城恒、太子坟、保子关等旅游景点。

Market Positioning 市场定位

Taking post-disaster reconstruction as the principal thing, reconstruction and local development strategy are combined. To pay attention to promote local features and give priority to dwelling rather than industry will make Weizhou a land of happiness for local people and ideal shopping and tourist place for visitors.

项目以灾后重建工作为主，将灾后重建与地方发展战略相结合，注重发展地方特色，注重先住宅后工业、先民生后产业，将威州打造成为当地居民居住的乐土和外来游客们购物、旅游、感受羌族风情的理想去处。

规划设计特色
Planning & Design Features

Street Planning 街区规划

After Wenchuan Great Earthquake, Weizhou Town was provided with assistance in construction by Guangdong Province. A general planning was made to reduce the planning population and scale of land use of Wenchuan County, change Qipangou area into residential, storage and logistics place from an industrial zone, and no longer rebuilt polluting enterprises.

The near-term target is to restore and reconstruct Weizhou to provide a solid foundation for social sustainable development. Target in promoting period is to further complete and improve the comprehensive development ability of Weizhou Town, optimize space layout and enhance the service and traction function of Weizhou. The promoting period target is rooted in local facts to develop Weizhou into a tourist town of Western Sichuan.

In order to create strong local features, landscapes of Weizhou are systematically planned. The general landscape planning constructs a structure of "one gallery and three zones". This structure means the Minjiang River and the Binjiang River connecting Qipangou scenic zone, center scenic zone and Yanmen scenic zone to form rich and varied scenery. Besides, the scenic zones are planned separately. Qipangou scenic zone mainly presents industry and living scenery; center scenic zone shows comprehensive living scenery; and Yanmen scenic zone shows education and living scenery.

汶川地震后，威州镇由广东省进行对口援建，广东省城乡规划设计研究院和广州市城市勘测设计研究院在《汶川县城总体规划（2005-2020）》的基础上编制《汶川县城（威州镇）灾后重建总体规划（2008-2020）》，此次规划的变化主要在于：压缩汶川县城规划人口和用地规模，中心城区部分人口向雁门坝、七盘沟片区分散布局；将七盘沟片区由县城工业小区规划为居住、仓储、物流集散地；污染型企业不再恢复重建。

规划的近期目标为完成威州恢复重建任务，为经济社会可持续发展奠定坚实基础。发展提升期的目标是在恢复重建期基础上，进一步完善和提升威州镇及县城的综合发展能力，优化镇村布局和城镇空间布局结构，吸收生产要素向城镇集中，提升威州的服务和牵引功能，从本地实际出发，扬长避短，形成特色，把威州发展成为阿坝州域内交通便捷、环境优美、具有浓郁地方特色的川西旅游型城镇。

为打造出具有浓郁地方特色的川西旅游城镇，设计单位对威州景观进行了系统的规划。在总体的景观规划上，构建了"一廊三区"的总体景观结构，既以岷江滨江为景观廊串联七盘沟景观区、中心景观区和雁门景观区，形成滨江丰富多样的景观风貌。另外对景观进行分区规划，七盘沟景观区主要展现工业和生活景观，形成"十"字形景观构架，连接多个人文或自然景观节点；中心景观区主要展示综合生活景观，构建"一带两轴三核"的景观结构；雁门景观区以展现教育和生活景观为主。

Street Design Features 街区设计特色

Weizhou is area where residents are mainly the Qiang people. Therefore, its design concept is based on strong local and ethnic features of Weizhou. Xiqiang Culture Street and Guozhuang Square of Qiang style are built along the Minjiang River. Sepia interspersing with sandy beige is applied; sheep totems are embedded on walls; columns with sheep totems are stood on the street; and traditional watchtowers are adopted.

In addition, a large number of earthquake proofing techniques are applied on building design. For example, shock isolation bases are used in three schools to make their earthquake proofing degrees reach 8 degree or above, and meanwhile, ensure the normal connection of various function spaces.

威州是一个以羌族为居民主体的区域，因而在设计时，广州援建者从威州浓郁的地域、民族特色出发，着力建设现代羌城，将"西羌圣城"的理念贯穿于设计的全过程。在岷江一线建设具有羌族建筑风格的西羌文化街和锅庄广场。建筑色彩采用羌族喜爱的片石浅褐色点缀原木棕褐色，建筑墙上嵌入羊图腾，街道上立着羊图腾的柱子，另外还运用碉楼这一传统羌族建筑元素。

另外，建筑设计大量采用先进的隔震、抗震技术。所建的3所学校的主体建筑都进行了隔震设计，建筑与地面、建筑与建筑之间的构造做了细部设计，采用了隔震支座，使防震烈度达到8度以上，在满足隔震的要求下，保证建筑各功能空间的正常联系。

Modern Imitated Commercial Streets of Ancient Style

文化设施 Cultural Facilities

Wenchuan Museum 汶川博物馆

The main body of Wenchuan Museum adopts frame-shear wall structure with earthquake proofing degree of 8. It is composed of a exhibition hall, a culture hall, a library and a book center.

汶川博物馆主体采用框架剪力墙结构，按8度抗震设防，由博物馆、文化馆、图书馆及购书中心组成，建筑面积8 633平方米，由广州市对口援建，何镜堂院士主持方案设计。